CAMBRIDGE LIBRARY COLLECTION

Books of enduring scholarly value

Astronomy

From ancient times, humans have tried to understand the workings of the world around them. The roots of modern physical science go back to the very earliest mechanical devices such as levers and rollers, the mixing of paints and dyes, and the importance of the heavenly bodies in early religious observance and navigation. The physical sciences as we know them today began to emerge as independent academic subjects during the early modern period, in the work of Newton and other 'natural philosophers', and numerous sub-disciplines developed during the centuries that followed. This part of the Cambridge Library Collection is devoted to landmark publications in this area which will be of interest to historians of science concerned with individual scientists, particular discoveries, and advances in scientific method, or with the establishment and development of scientific institutions around the world.

In Pursuit of a Shadow

The title page calls the author of this 1888 work 'A Lady Astronomer'. She was Elizabeth Brown (1830–99), and the shadow she was pursuing was the eclipse of the sun on 19 August 1887, which could be best observed in northern Russia. Brought up by her father to make weather observations and to use a telescope, she became a member of the Liverpool Astronomical Society – on behalf of which she undertook her Russian expedition – and was later active in founding the British Astronomical Association. (The Royal Astronomical Society did not at this point admit women.) The book describes her journey, from her arrival at Hull to meet her travelling companion, to Russia, and home again. The actual viewing of the eclipse, at Kineshma, 200 miles north-east of Moscow, was spoiled by cloud cover, but her lively and observant account of her adventures is a fascinating record by a pioneering female scientist.

Cambridge University Press has long been a pioneer in the reissuing of out-of-print titles from its own backlist, producing digital reprints of books that are still sought after by scholars and students but could not be reprinted economically using traditional technology. The Cambridge Library Collection extends this activity to a wider range of books which are still of importance to researchers and professionals, either for the source material they contain, or as landmarks in the history of their academic discipline.

Drawing from the world-renowned collections in the Cambridge University Library and other partner libraries, and guided by the advice of experts in each subject area, Cambridge University Press is using state-of-the-art scanning machines in its own Printing House to capture the content of each book selected for inclusion. The files are processed to give a consistently clear, crisp image, and the books finished to the high quality standard for which the Press is recognised around the world. The latest print-on-demand technology ensures that the books will remain available indefinitely, and that orders for single or multiple copies can quickly be supplied.

The Cambridge Library Collection brings back to life books of enduring scholarly value (including out-of-copyright works originally issued by other publishers) across a wide range of disciplines in the humanities and social sciences and in science and technology.

In Pursuit of a Shadow

By a Lady Astronomer

ELIZABETH BROWN

CAMBRIDGE
UNIVERSITY PRESS

CAMBRIDGE
UNIVERSITY PRESS

University Printing House, Cambridge, CB2 8BS, United Kingdom

Cambridge University Press is part of the University of Cambridge.
It furthers the University's mission by disseminating knowledge in the pursuit of
education, learning and research at the highest international levels of excellence.

www.cambridge.org
Information on this title: www.cambridge.org/9781108074445

This edition first published 1888
This digitally printed version 2014

ISBN 978-1-108-07444-5 Paperback

In Pursuit of a Shadow

BY A

LADY ASTRONOMER

JOHN BELLOWS
STEAM PRESS, GLOUCESTER
154475

IN PURSUIT OF A SHADOW

CHAPTER I

HULL TO CHRISTIANIA

It was about 4 p.m. in the afternoon of July 15, 1887, that I alighted from the Midland express at Hull and met my travelling companion. I had left my home on the previous day and had passed the night at Cheltenham to enable me to take the 9.20 a.m. train and all day I had been travelling North with only one change of carriage at the Milford Junction.

My companion, whom I will call L, lived in one of the Northern Counties and we were bound for Russia to see the great Solar eclipse of August 19, so that it may be imagined we met under circumstances of pleasurable excitement not unmingled with awe at the prospect of countries strange and unknown and of people of unknown tongues among whom we, with but slender linguistic acquirements, were about to

venture, nor had our friends and acquaintance
spared us the usual amount of alarming pro-
phecies, first and foremost among which stood
the terrors of Russian Custom-houses, stern
officials and a perpetual background of police
supervision.

Our packing had been accomplished in the
teeth of numberless hampering counsels. Books
had been interdicted by one friend, sketching
materials viewed doubtfully by another, while a
third anxiously specified that nothing was more
important than the loose arrangement of all
corners in our trunks in preparation for invading
hands; added to this my beautiful little telescope,
the loan of an astronomical friend, in its oblong
black box, certainly presented a suspicious
appearance almost suggesting dynamite! However for good or for ill our belongings were
packed up and we were ready to start at last. L.
did not join me for the first time as travelling
companion; she and I had journeyed over sea
and land together before, and I had proved her
value in that capacity to my complete satisfaction, and our dinner at the Royal Station
Hotel was a very cheerful and inspiriting one
in the company of some of her own people who
had come to start her on her travels.

The evening was not very propitious. Heavy
rain fell and we heard thunder about 6.30 but
it had cleared by the time we went on board
our ship, the Rollo, of the well-known Wilson

line, at 9.45. All was calm and promising then and the inspection of our berths which rather to our discontent were in the ladies' saloon proved encouraging. There was not the bustle we had dreaded, and we were able to prepare for the night without disturbance when we wished. The steamer was not due to start till 12, and we therefore had to make our good-byes long before but L. could not make up her mind to the final parting so early and merely lay down without undressing, going on deck later to have the last glimpse of her brother and sister who had come down to the docks with this intent, and as the Rollo, leaving her moorings about 2, slowly passed along, their last farewells were accomplished by touching umbrellas from shore and deck. But though I remained below sleep was not possible for a long time, and I thought the tramping overhead, the horn blowing and windlass winding would never cease.

Save for this, the soft electric light in the cabin, free from oily odours, would have been most soothing, the berths were comfortable and the stewardess perfect in her way. But even the hubbub of a steamer ceases at last, and it was a pleasant surprise when the arrival of a delicious cup of 7 o'clock tea and rusks, an unexpected luxury, proved to us that our first night out of England was well over. Breakfast was, or was supposed to be at 9. Our fellow passengers, numbering 45 in the saloon, were

chiefly bound for Norway, and our companionship was thus for too short a time to admit of
much intimacy.   It was a fine morning, and
when breakfast was over we sat on deck enjoying the cool air after a hot night.   We were now
in open sea, no land visible, the water a lovely
blue, only some pillary thunder clouds presented
rather a threatening appearance.   We watched
some gentlemen playing shuffle-board, and
afterwards accepted their invitation to join in
the game, proving that our heads and hands
were thus far steady enough for a little whole-
some exercise, but, alas, this happy state did not
continue long, the wind freshened till the sea
became quite rough, and though we kept up a
brave front till after lunch it was not possible to
do so much longer or indeed to find shelter on
deck, and the rest of the day and of the night
following was spent in great discomfort in our
berths, nor were we able to make our appearance on deck until one o'clock the next day, and
probably courage would have failed us then but
that we were due at Christiansand at 1.30 and
were anxious for our first glimpse of Norway.
Unfortunately it was then raining steadily so
that there was no temptation to land, nor
were the wet decks conducive to comfort on
board.   We were moored close to the shore and
our first sight of a Scandinavian town reminded
us of Canada.   We had a view of wooden
houses, mostly warehouses, and there was a

church not far off, but the cessation of motion
was, it must be confessed, more welcome than
the loveliest view would have been, and indeed
it so raised my companion's spirits that before
long she summoned courage for a short walk,
though without, as far as I can remember, bringing back any very definite impressions of her
first touch of Norwegian soil. The day, be it
remembered, was Sunday and on the return of
the errant passengers the Captain read prayers
in the saloon, a shortened form adapted from our
prayer-book, for the Wilson line, with several
hymns, and by this time we were sufficiently
recovered to enjoy it, and I think a religious
service is never more restful and soothing than
amid the turmoils and changes of travel.

We reached Christiania about 9 on Monday
morning, (July 18th). The land when first
sighted looked flat and uninteresting, but as
we drew nearer, the long low outlines seemed
to rise higher and higher, rocky prominences
crowned with stunted spruce firs came into
view, soft grey hills filled up the background,
lovely in their misty faintness, and then, en-
circled by these fair heights, with its church
spires, wooden chalêts and large warehouses
many coloured and of lovely tones of dull green
and ochre, the city of the sweet sounding name
revealed itself. We could but feast our eyes on
its beauty. Many of the public buildings were
embedded in trees, and amongst them the king's

summer palace, white with turreted towers, was conspicuous, while every movement of our propeller seemed to reveal some new and striking conformation of the girdling hills.

Then came the landing which was to have been our first experience of custom - house severities, but to our surprise the luggage was passed with hardly any examination, my own box not being even opened, so choosing one of the carriages which were in waiting, we were quickly driven to the Grand Hotel.

How soon the charm of a new land asserts itself! The most ordinary bed-room in a foreign hotel has something distinctive, and therefore pleasing even to a hardened voyager. Our eyes rested with satisfaction on the little characteristic signs of Norwegian life and habits, but our chambermaid was German, as proved to be very usually the case, even in Russia.

Having thus housed our boxes we felt we must lose no time in seeing a little of the town, so with the help of the hotel porter engaging a carriage we took a short drive. Passing through paved streets and clean squares, and by many public buildings, among which the king's palace and the Parliament house were conspicuous, we reached the University which is open three times a week, and there in an adjoining building, saw that most interesting old relic the Viking Ship. Little did I think when many years ago I read the graphic description of its discovery

in the *Times*, that my eyes would ever actually
see the original—the aged vessel which had lain
for centuries embedded in the sands of the
Baltic. So carefully was it exhumed at the
time, that it has been retained in almost perfect
preservation, and here we saw it lying in the
house which was erected for its shelter, its
ancient timbers blackened by time and by its
long continuance under water, but still holding
together, though it seems a marvel that they
should do so. The keel of the ship is very
sharp and almost perfect, and the frame-work
of the great ribs so little displaced that the
original form and proportions are discernible at
a glance. It stands on a raised platform protected by railings, and the building also contains
the actual bones of the old Scandinavian
Monarch, who, according to the custom of his
time, was at his death buried in his vessel,
which was then sunk beneath the waters for his
final rest. The bones are preserved in a glass
case in a corner of the room, but the ornaments
and other accessories of kingly state which were
also found were removed to the Museum. This
sight was the most thrilling incident of our
expedition, and immediately after we returned
to our hotel for dinner. The fare was good, but
the dish which chiefly remains in my memory
was one containing no less than ten different
kinds of fruit, fresh and dried. In the afternoon we sallied forth again, and this time we

took a steam ferry to Frederiksbor on the other side of the harbour, where are villas and shrubberies, seats and landing places, so that an hour passed pleasantly. We made our first sketches, and then went back again by ferry to tea, and I may mention that by this time we had completely fallen in love with the Norwegian horses, or rather ponies, for they are small, but stout, strong little animals, often indeed very usually cream coloured, with long manes and tails, and a most friendly Christian expression of countenance, indicating kindly usage and gentle treatment. A curious habit obtains among the drivers of country vehicles, market carts and such like. When leaving their carriages, or when stopping for any length of time they make all secure by tying one leg of the horse, for which purpose a long strap or rope is affixed to one of the shafts with a leathern loop at the end, which is buckled round the animals leg, just below the hock. I watched a woman doing it with the greatest celerity, and thought it a comical illustration of our national adage "tied by the leg" which might be brought forward as a philological proof of our Scandinavian origin. Of these good little animals we made personal trial the next day (Tuesday, July 19) to which I must now pass, for on applying to our obliging "maitre d'hotel" for advice he suggested a drive to Frognersæter, a summer residence of Consul Heftog, whither we started in a carriage and pair after our breakfast. On leaving the city, the road passes through open fields with many flowers by the road sides, among which we noticed hare-bells, yellow bedstraw, white chrysan-themums, and one species of bright coloured geranium, and then it winds upwards for miles through spruce forests. It was difficult to believe we were not in Switzerland, the cool damp woods and pine needled paths recalled many old ex-periences, but the sturdy little horses, who made nothing of the long ascent, were of a different type, and so was our driver, who spoke only Swedish. Arrived at our destination, we found it to be merely a summer chalêt in an open space with others scattered about. Its beauty consisted in the lovely view of the dis-tant city and sea stretched out far below us, the exquisite colouring of which, with the ever changing reflections of the summer clouds, made a picture never to be forgotten, though one impossible to paint. We lingered here for about an hour, children were gathering whortle-berries, and we found the leaves of the hepatica, but there was too much shade for any wealth of flowers. Presently L. noticed our driver give a significant tap to the shafts of our carriage, at which hint we again took our seats, and reaching our hotel at 1.30 found the nominal " table d'hote," which is really a common opportunity for private dinners, which are ordered from the "carte" going on in the "sal speise."

In the afternoon we took the steamer to Oscarshall, the king's summer palace, but as this excursion had no very distinctive features it need hardly be described in detail. It gave us an opportunity of seeing more of the people of the country than our morning drive, the steamers being crowded with afternoon excursionists. The poorer women invariably wear a light handkerchief tied over their heads, and they are all clean and well-behaved. I much enjoyed a quiet rest on the shore, attempting to catch some of the soft distant effects in my sketch book, leaving L., who is a far better sight-seer than I am, to "do" the palace. And here I may confess once for all, that but for her greater zeal and gentle but persistent instigation, much that I did see would have remained unseen, for I am country born and bred, and the love of quietness inherent in such, suffers much strain and self-denial in enduring the life of towns.

On retiring to rest at night we had some difficulty in explaining to our chamber-maid the exact time we wished to be called the next day, until L. cut the knot by a vigorous rapping on the door and by risking the well-being of her watch by twirling the hands round to the desired hour, a process which soon enlightened the bright little woman as to our wishes.

The daylight was long in going, and we could read fairly large print without a candle till nearly 12, proof positive that we were indeed in the land of the midnight sun.

CHAPTER II.

CHRISTIANIA TO STOCKHOLM.

We left Christiania, after an early breakfast, at 7.30, in a pretty little omnibus with blue silk curtains, and were glad to find our English gold readily accepted at the station.   The currency of Norway and Sweden is the same and we were already fairly familiar with it ; we were indeed fast becoming converts to the decimal system as far simpler and easier to remember than our own.

As we travelled first class we were alone all the way.   The carriages resembled our English ones till we passed the frontier at Charlottenberg, when they were superior in fittings and accommodation, with red velvet cushions to the seats and red curtains lined with white, the compartments opening one into another, their only disadvantage being that the windows, which are all fitted with double glass, were uncomfortably high.   We reached Charlottenberg at 12.30, and again our luggage was passed

without difficulty. Here we had our first experience of station refreshments, which were set
forth in a curious, not to say scattery manner,
on a long table, without order or arrangement,
the edibles being crowded together and flanked
by plates, knives, &c., which the passengers
appropriated at will, a fixed and moderate charge
being made. The coffee was excellent, with
delicious cream, and we found the slices of
bread, spread with cheese or meat, very convenient for a hasty repast. L. was by this time
becoming very clever in utilizing her few
recently and rapidly acquired words of Norse
or Swedish, so that I was relieved from all
anxiety on that head, especially as I saw how
the quick flash of her animated eye and gestures
emphasized her speech, almost forcing a respon-
sive comprehension from the listener.

During this day's journey we saw no towns.
The little wooden stations might, to all appearance, have been the only places of habitation.
We passed through miles and miles of pine
woods, often crossing rivers and seeing lakes,
lovely with white water lilies, but no hedges or
fields, all being broken undulating ground. After
reaching Sweden the country was more burnt,
and we noticed small fields of oats or a few
patches of wheat, ripe, but not cut, while in
many places hay-making was still going on, the
hay being dried on long hurdles. There were
but few cattle to be seen. The railway banks

showed the same flowers we had seen at Christiania, and also an abundance of the tall rose-
bay willow herb.   The soil, from the grey rocks,
appeared to be the mountain limestone.   The
chief feature in both countries was still pine
woods—spruce firs on undulating slopes as far
as the eye could reach, their sweet resinous
odour penetrating the carriages as we passed.
It was an enjoyable day with but little fatigue,
though the eye does at times get tired of the
incessant watching demanded by a new country,
when it seems a waste of precious opportunities
to indulge in a nap, even though the objects
seen are, as a rule, monotonously alike.

We reached Carlstadt at 4.45 p.m., and found
the Stadt Hotel, where we had decided to go,
within walking distance, our luggage being jolted
over the wide roughly paved streets on a horse
truck.   The hotel was quite Swedish, no other
language being spoken, with large carpetless
rooms and white porcelain stoves in every room
reaching to the ceiling.   Our bedroom had rather
the effect of a private ward in a hospital, with
its two little narrow beds covered with a sort
of regulation quilt, and provided with one thin
striped blanket.   The sloping bolster and square
pillows certainly were not like London, nor was
the little wash-stand, with two tiny basins, made
to shut up, and, when closed, looking like a
table.   We decided to have tea, and were
served with good bread of different kinds, also

biscuits resembling pastry, and there were cakes of a black porous bread, hard and leathery, on a separate plate, which proved afterwards to be a universal accompaniment of such a meal, though (not to our surprise) rarely partaken of.

Our landlord was sleek and fat, and our landlady more than fat—words fail to describe her ample dimensions; the waiters were all girls, and we found them quick in divining our wishes, but no one seemed able to explain to us, or indeed to know, when the post went out, though L. felt certain she had put the question in the most lucid manner, beside pointing to the clock and to the letter-box. This question being of necessity put aside, we started for a walk, and, after crossing the broad wooden bridge, followed the course of the river Ting-valla. It was a lovely evening, women were washing clothes in the river, beating the linen with mallets on boards under a shed, as one sees on the banks of the Seine in Paris. Pretty wooden houses, often with balconies and sheltered by drooping trees, were dotted all along the road. They had flower-beds, but no gravel walks. We noticed many of our old-fashioned English annuals, clarkia, nemophila, and a small white star-shaped flower, associated with my childish days, and in the distance we could see the mowers still busy in the evening light, while close at hand every one seemed sitting out of doors enjoying the cool soft air.

2

As I was sketching on the river bank, rather in
terror of mosquitos, two beautiful cats came
prowling along, looking very homeish and
similar to their English sisters, though with
rather more pointed heads, and tails more
curved.   Later on, as we turned homewards, a
red sunset made the windows blaze like rubies,
and the reflections of the houses in the water
were of marvellous beauty.   We were loth
to retire for the night.   We felt, now fully
launched on our journey, that we were indeed
in a new land, and the spell of it was upon us
and held us captive ; we determined it should
not be a wasted experience, and that the
present should, day by day, lay up its store
of new impressions, vividly realized, not easily
forgotten ; and if in my own mind an ever
present anxiety as to the great object of my
journey would now and again intrude itself, it
was as yet too far off to be disturbing.   We
left the next morning at 11.30, with the prospect
of a long day in the train before us.   Evidently
our foreign nationality was very apparent, for
railway porters and others were always ready
in offering help, in taking tickets, seeing after
luggage, &c., even beyond our requirements.
We had frequent stoppages during the journey, and at Laxa, which we reached at 2.30,
half-an-hour was allowed for dinner, the "mat
sal" being arranged in the usual manner, and
well provided with edibles.   The bill of fare

included soup, mutton and beef, vegetables, small pancakes, stewed black currants, and a kind of junket, and there were small bottles of beer, like miniature hock bottles, besides coffee. The country maintained the same character. As we sped along, the line of rail cut through stretches of pine woods for miles and miles, occasionally varied by patches of birch and Scotch fir, and we passed a constant succession of lakes and rivers, so that pine woods and watery ways seemed again the one feature of the journey. The day proved showery, which was in one sense to our profit, as it caused a lovely variation of light and shadow, now deepening the bluish green of the woods into blackest indigo, now flashing on the water of lake or river, while for a long distance we were followed by a splendid rainbow, which spanned with its aerial arch the freshly moistened stretches of mossy woodland or the occasional patches of oats or wheat. The railway banks were rarely steep or much inclined, and it was a constant mystery to me that we saw nothing of any towns. We would stop at a station and see only a few wooden chalêts, nor did there seem to be any people who might inhabit these invisible towns whose existence remained a matter of faith. The country seemed as often as not peopled only by the hooded crow *(Corvus Cornix)*, and these birds we watched with great interest, as they are rarely if ever seen in

England. They are large birds, the size of our
common rook, with black and grey plumage,
and are allied to the carrion crow. They were
constantly to be seen stalking about, and the
first time we noticed them L., with great excitement, made frantic endeavours to gain some
information about them from our driver, with
no further result than obtaining an indifferent
nod of the head in reply ; he evidently con-
sidered these feathered people quite beneath
his interest. And so the day wore on, and as
we neared our goal the sunset flooded the sky
with golden light, while little rose-coloured
clouds floated above the horizon, changing
again to deep cold purple. The light was reflected in the lakes as we rushed past, and
wreaths of white vapour rose from the water
and streamed away into the distance. It was a
mysterious approach to Stockholm in the semi-
darkness. The evening star, seen for the first
time since leaving England, was glittering low
down in the sky, and the number of lagoons
surrounding the city which caught the last
colours of the sunset, and the twinkling of
scattered lights added greatly to the impres-
siveness of the scene. When actually in the
city, the streets, to our surprise, were not lighted
up, but in one place we saw public gardens
crowded with people and with rows of brilliant
lamps, these lights being also reflected in the
water. Omnibuses from the Grand Hotel were

at the station, though there was no hotel porter
on the platform, but, thanks to the system of
baggage tickets, we had no trouble, nor, as yet,
had heard anything of over weight, and we
flattered ourselves that the telescope-case had
rather an appearance of fishing apparatus,
which must be a very usual item among travellers in this country. The hotel is a very large
one, making up 400 beds, and we found the
"salle à manger," with its little round tables and
padded cane chairs, brilliantly lighted with
gas and much decorated. There was a good
deal of gilding, many mirrors, cut glass lustres,
and the great white stoves were pricked out
with gold. All this looked very fine by gas-
light, but by day the style of ornament was
wanting in taste, and not to be compared with
the best American hotels. It was ten o'clock
by the time we arrived, and after taking a little
bread and cheese and milk, we were glad to
retire to our comfortable bed-room, in " New
York Gatan" (for each landing has a separate
designation of this kind). After a rather late
breakfast the next morning (July 22) we made
an expedition to the Post Office, and found our
ever welcome letters awaiting us, and afterwards
commenced our exploration of the city, by taking
a steam launch to the King's palace. These
steam launches are a great characteristic of the
place. They are darting about, plying backwards and forwards every few minutes, and are

really only ferry boats, and quite distinct from
the larger class of steamers which run to the
islands or places of amusement on this fiord of
the Baltic.   They are very convenient, and the
charge for the passage is trifling.   A boy goes
round to collect the few *öre* demanded in pay-
ment, as a receipt for which he gives a small
piece of copper, which you drop into a box,
with a slit, provided for the purpose, as you
step on shore.

Before going to the Palace we took our
tickets at the booking-office for the coasting
steamer "Tornea," by which we had decided to
leave on the Monday following, and, for the
first time, had to give up our passport into the
captain's keeping.   We did not go over the
palace, merely passing through the court-yard,
and then walked through some of the old
narrow streets of the city, which are very
picturesque, with lofty houses, but without balconies.   From thence we reached the market,
and wound our way in and out of the numerous
stalls arranged under awnings.   It was a capital
place to see the country women and the produce
they offered for sale, which included almost
every kind of vegetable and fruit, beside the
flat netted baskets for the live chicken, which
excited our compassion by the way they
stretched their little feathered heads through
the netting, anxiously picking up their corn as
best they could.   The women all wore white
or coloured handkerchiefs, and their chattering

and chaffering in Swedish was amusing. We
came in for our share of notice, with, as we
supposed, occasional invitations to purchase,
and then we made our way back to the hotel to
dinner, a meal which seems very generally taken
early, two or three o'clock being a usual time.
We had soup, salmon with potatoes, roast beef,
and a delicious light pudding, with ice cream.
On the clean wide pavement outside the hotel are
tempting little seats, with benches or chairs,
arranged for parties of two or three each, under
a separate awning, and also shaded by huge
pots of bay trees.

It was delightful to sit there, and there were
subjects of all kinds for an ambitious sketcher.
Before us was the quay, with ships of all sorts and
sizes ; on the opposite bank tall houses tower-
ing up as a back ground, while further behind,
some domed structure reared itself. In front
of us passed a constant stream of people of no
specially national appearance, with the exception
of the handkerchiefs of the women. But the
one feature of Stockholm which we shall never
forget was its intense cleanliness, purity of air,
and absence of dust. The paved streets and
intersecting canals, or what might be called such
save for the greater width of the water which
really is a part of the sea, prevented the possibility of dust in this Swedish Venice. You
might, without exaggeration, sit down fearlessly
in any street, or on any bridge. It was a

refreshment merely to breathe in such clear
sweet air, and the people had the same appearance of cleanliness and freshness.   L. was
always saying " How deliciously clean !" "and
you know," she said, "we were to have been
smothered with dust from the time we landed
till our return."   The one occasion on which
we did encounter dust at Stockholm was in an
excursion which we made in the afternoon of
this day, and which was in itself a mistake, for
which our guide-book only was to blame.
" Let no traveller omit to visit the Djurgarten,"
said the book, so to the "Djurgarten" we obedi-
ently went by steamer.   We found it what may
be styled a tea-island, for the whole place was of
the tea-garden character.   There were avenues
of trees, *dusty* gravel paths, grass plots, a
theatre, endless benches, and unlimited tea and
coffee.   Certainly we ought to have ascended a
tower which must have commanded an extensive view, but this we omitted to do, so, after
watching the crowds of Stockholmers, all of
what we should call the lower middle class, till
we were tired, we returned home in moderate
spirits, and sufficiently fatigued.

Very different to this was an excursion which
we made to the island of Vaxholme on the
following day (Saturday, July 23rd,) a day
which stands out in our memories as a white one
in our journey.   It was not a brilliant day overhead, but rather cloudy and showery, awakening

our fears of a downpour, but this held off, happily, and the voyage, which occupied an hour and a half, gave us a constant succession of pretty views as we steamed down the wide *fiord* or arm of the Baltic. These banks were dotted with villa houses, with their verandas and gardens, not exactly after the orderly and cultivated style of our English gardens, but still looking tempting from the water, with winding paths, shrubs, flower-beds and summer houses, and always seats and tables under the trees, each with its little landing-stage and flight of steps down to the water, and often with its private bathing house. As we left Stockholm further behind, the shore became more rocky, and our beloved fir trees re-appeared. Pleasant paths wound upwards among them from the wooden houses, and then we reached Vaxholme. Its speciality is fishing and boating, and the wooden buildings appropriated to these occupations were very pretty, being painted in rich colours, indeed the colouring of the whole place was so bright and varied that I was constantly regretting that I had not brought my paint-box, as no pencil outline could possibly do it justice.

We strolled up the village for some distance, constantly coming upon the rocky foundations which jut out everywhere, and at every bend and turn there would be a flash of the clear water of the sea, a gleam of red or orange in a

coloured roof or wooden house, or the glint of
a white handkerchief on a woman's head—of
blue-green pine foliage or gayer garden; a sort of
scene that led one on and on till a sense of time
passing and no sketches accomplished, suddenly
awakened us to the fact that we had passed no
end of suitable rocky seats, and now must
do our work standing or borrow a chair. L.
started off to secure the latter for me, and I
followed more leisurely.   She had noticed a
little house of superior order a few paces off,
embowered in shrubs, with the universal round
table and wooden seats for out-door life, and
espied two women within, who had evidently
been watching our movements.   In a moment
L. had her little manual opened at a suitable
sentence, and preferred, as she imagined, the
request for "a chair," at the same time pointing
to her sketch book and pencil; but by the time
I had joined her the result of her request had
produced, instead of a seat, a peal of uncon-
trollable laughter, in which we could but join.
Once more L. pointed to her sentence, and once
more the merriment burst forth.   She had
touched the wrong line, and had asked for
"some roast beef!"   Then we pointed to a
stool, but our light-hearted friends could not
easily recover their gravity; however we
did manage to make our want known at last
and carried off a chair, only, unfortunately,
when they enquired *where* we wanted it, we

answered "at Stockholm," which reduced us all once more to our former condition!

These women were evidently of a superior class, and we discovered later that they had a room to let, and also that a doctor was to be consulted at their house. On our return they seemed anxious to further our education, and were most pressing in explaining the proper pronunciation of certain words, evidently seeing they had an apt pupil in L; but a man, whom we descried reading a paper in an inner room, fought shy of the strange ladies, and could not be induced by his women folk to come to their aid. We wondered whether he could be the doctor, but L. decided that to be impossible, so dumb and morose a physician being a thing not to be imagined. We altogether had quite a sociable time during this expedition, for we invaded another house, and saw a woman busy making "*pankakka*" at a stone fire-place, something like an oven, in the cleanest of little kitchens, opening out of which we had a glimpse of a tiny sitting and bed-room all in one, in perfect order, with bright red geraniums in one window, and a little sacred picture on the wall. Pancakes suggested dinner, and we made our way back to the hotel, a queer rambling erection, all galleries and outside eating-places. Our table was in a corner overlooking the water, boats were constantly passing, gulls and sea swallows skimmed the air with shrill cry, the

river steamers came and went, and at last it
was time for us also to embark, and indeed this
long day in the air was causing us both an
overpowering sleepiness, moreover rain came
on, and our voyage home was a period of welcome quiet.   Tea also was welcome on our
arrival, and then, to finish this lovely day, we
ascended by the lift to the top of the hotel,
whence, by a flight of steps, we emerged into an
open place with seats and the inevitable round
table.   Such a glorious place for seeing the
stars! But it was the sunset we had climbed up
to see, and we watched the red and lurid light
which touched the city or the shipping as with
fire, while far below, the bustle of a parting
steamer reached us in softened tones ; then a
burst of softer singing succeeded the clamour
of voices, handkerchiefs were waved, the pad-
dle wheels revolved and bore them away.   We
descended from our post of observation, and
this delightful day was over.

CHAPTER III.

STOCKHOLM

It was pleasant to wake to a Sunday in clean Stockholm with its pure clear air, for the sense of outward purity and holy rest seem naturally united.

We made choice of *"Santa Clara"* as our church for morning service, a Swedish Lutheran church, and arriving there somewhat before the hour, we sat down outside in the cool shade of some trees to wait, with grass of a refreshing greenness before our eyes. There were a few people, chiefly women, also waiting, and a deep and sweet-toned bell presently clanged out over our heads. The church was hardly half full when the whole congregation had assembled. The interior of this large building was plain, but there was an altar-piece and a choir as in Roman Catholic churches, and we were surprised to see the officiating "minister"—for a lady, of whom we had asked the way, assured us there was "Keiner *Priest*"—arrayed in gorgeous vestments, which however were only assumed for the early part of the service, the "Geneva gown"

appearing when he mounted the pulpit. The
sermon was short, occupying only about fifteen
minutes, but it was succeeded by a very pro-
longed reading from the Epistles. The strain
of listening to this in a foreign language caused
my attention to wander to another book, that
of the foreign people there present, and I could
not but think how the points of resemblance
must far outnumber those of difference in different lands and lives. The sorrows that touch,
the comforts that heal, the rapid passing of the
familiar days, the rapid nearing of the days that
shall be as familiar, though we trust infinitely
more blessed.

In the evening we walked along the quay and
found a good seat on some wooden benches,
sheltered by shrubs which formed a partial
screen from the passers-by, while in front was
the water, now glittering in the sunset light. It
was evidently a favourite resort. Our bench
was shared by two women apparently of the
shopkeeping class, and, attracted by their pleasant faces, we fell into conversation, we ourselves
being evidently objects of interest as strangers.
To talk in Swedish was not possible, but
they knew a little French and German, the
former language being the one in which I was
most at home, while L., of the two, spoke
German more fluently, and thus our talk was
an amusing mixture of both. It was a little
foretaste of pleasant polyglot conversations, as

yet far ahead of us in the heart of Russia. The women were greatly interested, and answered our questions respecting the winter, the king, etc., with much animation, and in this duplex tongue. It was like tossing two balls backwards and forwards. *"Vraiment"* said I, *"Wahrlich"* echoed L., *"Ach Ya,"* cried one woman, *" Mais oui certainement,"* replied the other! And then we said a cordial "Good-bye" and walked slowly back. The public gardens were crowded with people, the throng indeed seemed to thicken as the evening advanced and the darkness deepened, for it was nearly dark by the time we reached our temporary home and haven of rest, rest not unneeded or unwelcome.

Monday was devoted to a visit to Upsala, a journey of an hour and a half, through similar country to that which we had already passed. Haymaking was going on ; long waggons drawn by oxen carrying the hay. It was another lovely day, without great heat. We first visited the Cathedral—a large building, approached by a flight of steps, and now in process of restoration, part of the exterior being almost concealed by scaffolding. Our driver hunted up the woman-sacristan, and we were fortunate, when admitted, in joining company with a party who were going round under the guidance of a man able to explain to us the tombs, monuments, and other objects of interest. The church contained a series of

pictures representing the life of Gustavus Vasa,
and our guide also showed us the tomb of
Linnæus—a flat slab on the ground.   It was
covered with the dust of the restoration, which
he had to sweep away with his hand before we
could see the grave and read the name of the
world-famed botanist, with the date of death
and burial.   There was also an iron case containing regalia, so that it seemed a curious
combination of church and museum.   A figure
of our Lord stood over the altar, but there was
no coloured glass.

Leaving the Cathedral, we proceeded to the
University.   The session was over, and the
place appeared closed and deserted, but, after a
long knocking, a gentlemanly man, whom we
concluded to be one of the officials, appeared,
who very courteously showed us round.   The
entrance-hall is very fine, with a domed roof,
and the lecture and class-rooms, concert and
senate-halls are spacious and complete, all
painted in soft dark colours, and with handsome
oak chairs covered with dark blue cloth, while
the walls were appropriately hung with portraits
of scientific celebrities.   We then asked for the
Observatory, and our conductor sent a man with
us to show us the way.

It was at some distance, and we had to cross
a court covered with rough grass, and traverse
some wide cobble-stoned streets, before reaching
the building, which is domed, and stands by
itself in an open space.

Mr Hildebrandtson, the resident astronomer, was absent, but his son received us very kindly, and took us round, and into the instrument rooms, one being especially devoted to transit instruments and also containing a 9-inch refractor. In another was a splendid photometer on which I recognized the name of a well-known English firm. " Troughton and Simms." Our guide, who only knew a little German, spoke very modestly of the place, remarking, when I asked if they would send an observer to the eclipse, *"Nein, es ist nur ein kleines Observatorium."* The Meteorological Institute is close by. It has come a good deal into notice of late years through a series of investigations on the motion and height of clouds, which have been carried on under the direction of Professor Hildebrandtson, and for which a beautiful instrument was specially designed by Professor Mohn, of Christiania, particulars of which may be found in an interesting paper, by the Hon. Ralph Abercromby, in *Nature* (August 4th, 1887). The results of this work, which, on suitable days, is carried on thrice in the 24 hours, are watched with great hope and interest by the scientific world, and bid fair to invest Upsala with a reputation unique of its kind. It is not a government Institute, and is entirely maintained by the University, and we were interested in hearing that beside Messrs Ekholm and Hagström, the assistants, a lady is employed in the

3

telegraphic work and in making some of the computations. Upsala seems to be the centre of intellectual life in Sweden, and through the purity of the air it must be admirably calculated for astronomical work. These interesting visits being over we went to the cemetery, which is a large one, and then searched for some hotel, or restaurant for refreshment, but half the place seemed asleep, and the houses closed, owing probably to the absence of the students, who are its chief vital element, and who only reside here during the winter session. At last we did find a place where the "*mid-dag*" was going on, and obtained fish and boiled lamb, and then, to rest our weary feet, we took an omnibus to the station, but had a long waiting there before the train started.

This was the only occasion on which we travelled 2nd class, but we had, as at other times, a compartment to ourselves, and found the carriages comfortable and clean, with cushioned seats and a notice "*Röknung ist forbjüden*," (no smoking allowed).

In the evening we had heavy rain with thunder. We watched the storm from our window, and also watched with interest a most persevering woman, who was employed in throwing chunks of wood from a great heap opposite our hotel, into the cellar of the house close by, and who worked steadily on till the

last was disposed of, as though perfectly uncon-
scious of the torrents of rain which must have
drenched her clothing.

We left Stockholm the next day in the
evening light of a clear sky, the thunder-storm
of the night before having apparently cleared
the air.

It was nearly 6.30 when our vessel, the
Tornea, glided from the quay, and the clinging
haze and more pronounced shadows of the evening touched every object with a peculiar charm.
The fine building of the Museum stood out in
bold relief, and our hotel, the Grand, with its
extensive front, and white outside-blinds flutter-
ing in the breeze, was a conspicuous object ; and
there was a long, low bridge with wide arches,
which was a perfect study of softest light and
shade. Even the steamers, which were everywhere, added to the effect of the scene, their
red funnels making a point of colour against the
grey and stone-coloured buildings and quays.
All passed before us as a shifting scene, and
short as had been our stay, there was a feeling
of a last good-bye not unmixed with regret; only
in the rapid changes of a journey new interests,
and quickly-born affections spring up in such
swift succession, that the pain has not time to
become developed, while the pleasure remains,
safely stowed away in the recesses of the
memory.

Less than an hour brought us to little Vax-
holm, with the evening brightness radiant on its
Indian-red fishing-houses, on its fort on the one
hand and on its green bastion, bristling with
guns, on the other. Voices from the hotel cheered
us as we passed, and our band struck up merrily
in reply. Then the sunset colours deepened,
touching the waters with transient glory. Long
straight lines of cloud barred the western sky;
beneath them stretched a bay of softest tender
green; and presently the perfect orb of the sun,
emerging for a moment from a cloud-bank, sank
behind the fir-belted islands which stood out in
deepest indigo, black against the light. Later
on the clouds turned to crimson, then to a pale
orange, and then the moon, now fast sinking,
showed her silvery arc, which appeared and
disappeared among the bars of cloud. These
lovely colours lingered until long after 10, and
at 3.20 I woke to find the sun already risen, so
that the dominion of darkness was feeble indeed
on this wonderful night. At breakfast the next
morning (Wednesday, July 27) we first saw tea
in tumblers, which after the Russian fashion is
taken without milk, with slices of lemon and
sugar. The breakfast, which was excellent,
including hot cutlets with vegetables, was laid
on deck, a great improvement on the scrambling
supper of the previous evening in the hot and
crowded saloon. We were constantly passing
low rocky islands, mostly bare with grey or

sand-coloured rocks, many with little piles of pebbles, painted white, erected doubtless as land-marks. Before noon we reached Hango, with its small hamlet of wooden houses, and then more islands, and yet more, pine-clothed or barren, but all picturesque. The sea was almost calm, and the white light of the sky reflected in it gave it a beautiful glassy appearance, and in this glassy sheet of light were set these innumerable islets, light and dark, grey, purple, green, violet—those farthest away showing only as dark lines in the water. So numerous were they, and some of them so near, that it really seemed at times as if we must touch them and run aground, but as one almost waited for the concussion, a turn of the vessel would send us rapidly past, often only to find another on our bows, which, in its turn, was in a few moments left behind. Beautiful as it all was, it seemed to be a barren region; we saw neither bird nor beast, and rarely any human dwellings.

Dinner, also on deck, was a tedious process as there were 30 or 40 passengers to serve, and only 2 or 3 girls as waiters, so that the delicious strawberries for dessert were doubly welcome as announcing the end.

We got into more open water in the afternoon, and reached Helsingfors about 6.30, the long line of houses being visible for some time before. The coast was still flat and rocky, but with more vegetation and more signs of habitation, such as hamlets and scattered houses, on passing which our band would strike up a lively measure. We were glad to land for a short time, if only to escape for an hour the continual smoking of our fellow-passengers, to me always a drawback to the comfort of a voyage, also we wished, for once, to set our feet on the shores of Finland. It was not a very delectable experience—everything looked dry, dirty, and dusty, nor was our first sight of a row of Russian droskies which were drawn up on the pier, inducive to the delights of a drive. Some American fellow-passengers who passed us at full speed in two of them seemed in imminent danger of contusions and catastrophies. Later on we found this extremely perilous appearance was less so in reality, but they are certainly a style of vehicle which require a training in steadiness of nerve, and as L. remarked, "only near and dear friends could share one comfortably" seeing one has at times to cling closely to one's co-passenger, the space allotted for the seat being of the smallest, and the ordinary rate of the free and joyous Russian horses a hand-gallop.

We found Helsingfors a larger place than we expected, numbering, according to Murray, 35,000 inhabitants, and now possessing the University formerly at Abo. The streets had nothing special in appearance, but the churches looked more Russian in their build and effect than the Swedish ones.

The market was being held on the wharf, passing through which L. and I walked up a hill covered with jutting rocks, which gave us a fine view of the town and harbour. We found a bench at the top and sat down for a time, then, returning to our ship, discovered to our relief that the noisy band and a large number of our passengers had landed for good. It was pleasant to sit quietly on deck watching the sunset, though it was not such a gorgeous spectacle as the last. The little steam-launches with their coloured lights, red on one side and green on the other, looking like marine glow-worms or fire-flies as the darkness came on, were plying here and there. The Finn women sitting in their white caps under the booths on the quay, had lighted their candles and were taking their tea, not, by the way, in glasses, but in big cups and saucers.

The town also began to twinkle with lights, and presently our Americans returned, and stood near us, chatting with us and with the captain. The gentleman announced himself as past 70, and told us he had made this journey 50 years before, and that it was all as fresh in his memory as though it were yesterday, adding that for a good first impression we ought to reach St. Petersburg in the evening. He was a pleasant old man, courteous and cultivated, but managed to upset L.'s gravity at dinner on one occasion, by gravely advising his daughter, who was drinking nothing, to "take a little moisture!"

It was 10.30 before we retired to our berths, a necessity we put off as long as possible, as the closeness below was apt to cause a slight return of our sea miseries, for though there was nautically no "sea" the Tornea did roll not a little at times.   When we awoke the next morning it was to the thrilling consciousness that before night we should be in Russia.   How the hours did drag that day!   There was little to look at—the coast was low and flat, we had left the islands behind, and the sun was blazing down on us in spite of the sea breeze.   Now it was so near, the prospect of adventuring ourselves, two unprotected females, into the giant arms of this great empire, looked more formidable than at a distance.   Our ignorance of the language gave a feeling of helplessness, and in Russia there are many drawbacks to the easy pursuit of common knowledge.   To begin with, the lingual characters are different, so that to see the name of a street is by no means to be able to read it, the pronunciation of the commonest word baffles one, and to add another element of confusion the old style still prevails, delaying the date twelve days, so that it became a positive mistake to imagine that we should land on the 28th of July.   That day was nearly a fortnight ahead.   We should land on the 16th; "And how," said L. "is it possible to arrange a journal in this state of things, or to account to one's own mind for a repetition of time actually lived through and done with?" " Moreover," I added, "the clock time will vary also in the *other* direction. The difference of longitude at St. Petersburg will cause a difference of two hours. 10 o'clock a.m. at Greenwich will very soon be 12 o'clock for us." "And what has become of those two hours?" said L, "where have we dropped them? Shall we ever in all our lives pick them up again?"

And every hour, with every pulsation of the engine, this strange land drew nearer—this mighty continent which was to be crossed by that line of shadow which had lured us away from little England in the hope of witnessing one of the grandest spectacles of the natural world.

It was about 2.30 when we noticed, rising in mid-channel in front of us, something which at first sight appeared to be an island—then a group of two or more islands—but which, as we approached, revealed itself as the mighty fortifications and granite walls of Cronstadt. Almost like islands of rock they were, rising in their rugged and motionless sternness sheer out of the water on either side, as though built by the Titans. Here was no architectural grace or adornment—no sign of human life—but, as we were close upon them, we could see the rows and rows of gun-holes, every aperture with its

deadly occupant.   This was at the entrance to
the harbour.   The harbour itself was crowded
with men-of-war, gun-boats, tugs, vessels of all
sorts.   A little apart from this warlike crowd
floated the Emperor's pleasure-steamer, and in
the far distance, on the main-land, Oranienbaum
and Peterhof were just visible, the latter place
containing the Emperor's summer residence,
where he was then staying.

After this it was not long, before—right in
front of us—out of the sea, or rather hanging
over it—appeared a long dark line, like a smoky
cloud, and in the midst of this cloud a spot of
light like a tiny glittering ball.   The cloud was
St. Petersburg, and the bright speck of light
was the golden dome of St. Isaac's!

CHAPTER IV.

ST. PETERSBURG

The nearer we approached St. Petersburg the more impressive was the scene, and especially after entering the wide canal, or waterway, which leads to the city.

After the grim impressions of Cronstadt it was refreshing to see peaceful haymaking going on on the banks, the workers habited in red blouses, and a sweet and flowery hay it must have been, judging from the fragrant scents wafted to us as we passed.

The city lies so low, from the fact of its being built on the sea level, that until close to it you can see little, then one gilded or coloured dome or spire after another, one of pale blue being very conspicuous, comes rapidly into view, while the great buildings and palaces seem to spring up suddenly close to the water's edge. There also we saw the new iron-clad, lately launched, with men still at work on her, and without her masts.

We were all standing at the prow in the blazing sunshine, which seemed to get hotter and hotter, and already, even before entering the city, the vastness, which is to me the great feature of St. Petersburg, as cleanliness was that of Stockholm, was unmistakably evident.

There was no confusion when the vessel stopped, and our belongings were in a very short time deposited on the wharf, where the custom house officers were already at work. Once more our dread of any trouble proved entirely fallacious—they only just opened our two large boxes, while the telescope, and even L.'s little bag of books, were passed without demur. We could almost fancy that the kind Russian consul in London, with whom I had had a personal interview, had sent secret intimation of our harmlessness, and recommended us to mercy. After this a *commissionaire* from the " Hotel de l'Europe" took us in hand, and conveyed us and our boxes thither in a dingy little omnibus. It is an immense place, with corridors running the whole length of the front, the rooms on the ground floor being partly occupied by shops. All the windows are double, with thick white blinds, which draw up and down like Venetian blinds.

Our large bed-room (No. 154) had four tables and a carpeted floor. The stair carpets were all covered by a white washing material, which attracted our attention by being covered with darns, so beautifully executed that they were really raised to the level of a work of art. The whole hotel had an effect of whiteness, except the dining saloon, which was shady and darkened, and grateful to the eyes in the sultry heat which we were now experiencing.

Only the head waiter and the *major domo* spoke English, but red-capped porters were always standing at the entrance, and carriages, with a pair of horses, constantly in readiness in the court-yard.

This evening we did nothing but rest and watch from our windows the busy life below. We were not in the *Nevski Prospect*, the Regent Street of St. Petersburg, but in a street leading out of it. Droskies were constantly passing, at their usual furious pace, and we noticed that the costume of their drivers, the full-skirted coats, flat, beaver hats, and enormous boots, which had been familiar to us in pictures from our childhood, remained unaltered.

The splendid equipages of the nobles we had little opportunity of seeing, as at this time of year they are, with few exceptions, out of town.

The next morning we resigned ourselves to the necessity of employing a *valêt de place*, and gave up our independence into the hands of one "John," a brisk, middle-aged little man, whose English was a singular compound of different languages. This personage engaged a carriage and pair, and, after transacting a little necessary business at the Bank and Post Office, we visited, under his guidance, the three chief

churches of the city.   This was the will and
arrangement of our guide, but was open to the
objection that by the end of the day the contents
of all the three were considerably mixed in our
minds, and needed much sifting and re-arrang-
ing.   And here I must remark, once for all,
that I can make no attempt at minute description of the details of the public buildings,
churches, and other sights we saw, to do which
would require a volume.   I can only endeavour
to chronicle the general effects which struck us
as we passed along, the impressions received
from the shifting visions of strange sights and
wonders which imprinted themselves on our
brains, to 'be re-produced in our thoughts and
fed on by our minds in after days.

As I have said before, I am a bad sight-seer.
To compile anything approaching to a guide-
book would be the last thing I should dream of
attempting, nor am I sure that the fund of
information that must be painfully heaped up
in order to effect this is a thing to be desired.
To come within the range of new and distinctive phases of life, to catch something of the
spirit of a new country, to carry away some
vivid pictures, whether of the wonders of man's
handiwork or of nature's rarer features, is, I
think, far more important—in a word, to possess
oneself of memory's picture-gallery, rather than
the collector's hand-book.   With this digression
I return to our morning's work.

We began by visiting St. Isaac's, the church of the golden dome. On first entering this church, on a week day, its immense proportions strike one with wonder, and this great size of itself produces a sense of softened colour and solemnity. Were it not for this, the over-powering quantity of ornament, of gilding and jewellery in the sacred pictures, and of precious materials as a coating or covering of the walls, pillars, etc., would cause an effect hardly short of tawdriness. In all Russian churches the pictures are literally encrusted with jewels. Our eyes wearied of the monotony of rich gems of all kinds, including diamonds of priceless splendour, with which these saints and martyrs blazed, the great peculiarity of the arrangement being that the faces and hands of the figures are left uncovered, flat on the canvas, while enclosed in a setting or nimbus of this raised and jewelled ornament.

This, we have been told, whether correctly or not I cannot say, has its distinct religious mean-ing, probably emphasizing the objection of the Greek Church to anything approaching an image. Everywhere are pictures, but no statues, the sacred "*Rizas*," or "*Icons*," are pictures of the Virgin and the Holy Child, and also of saints innumerable. As in Roman Catholic churches candles are to be bought within the building, and we constantly saw the poor people placing them reverently before the pictures. On entering

a church, especially on Sunday, these lighted
candles are to be seen shining everywhere in
their silver sockets.   They shone like groups
of stars in the dim distances of these vast
edifices, emblems of the undying life of the soul,
and under these lights sparkle and glitter the
gilded or jewelled settings of the faces or figures.
These, I should say, include, as well as Saints,
a great number of Emperors, who thus share
Divine honours, and doubtless, in the simple
minds of these untaught worshippers, in a
despotic country, the distinction between the
heavenly and earthly rulers is not strongly
marked.

In all Greek churches the *Iconostasis* is a
special feature.   It is the sanctuary, correspond-
ing with the chancel in our own.   It is a part
of the building at the east end, partitioned off
by a gorgeous frame or screen, covered with
pictures and emblazoned with gold, or, as in
the Kazan, cut out of solid silver, which reaches
to the roof; within is the holy place, where foot
of layman may not enter.   The Kazan Cathedral, which I cannot now describe, was our next
object, and then we went to the Church of
St. Peter and St. Paul.   This is smaller than
St. Isaac's, and chiefly remarkable for the tombs
of the Czars and Empresses.   Here Peter the
Great lies side by side with his two wives.   The
monuments are all large, oblong, massive structures, of pure white marble, with the names only

in gold letters at the sides. They would be really majestic in appearance but for the garniture of faded, and indeed often crumbling wreaths with which many are encumbered. The evergreens in pots placed round the encircling railings have a cool effect, but looked to us strangely unlike a church.

In all these buildings the spaciousness is more observable from the absence of seats. There are no forests of chairs, only a few benches round the outside walls, and therefore less possibility of a moment's rest to the jaded visitor. Outside this church stands the great fortress, and we could but think with a shudder of its many horrible dungeons. "John" also showed us the square where the Nihilist executions take place in the early morning.

After all this it was really necessary to take a little rest, but before parting company our guide arranged that we should go in the afternoon to the St. Alexander Nevsky Monastery, "to hear some good singing," which we accordingly did. The tomb of the Saint is of solid silver, and very wonderful, and the singing simply exquisite, though after a time it becomes monotonous. No instrumental music is allowed in the churches, and they spend large sums on good voices. It was the monks who were singing here, their long hair looking to us very unpriestly, as well as unbecoming. They all seemed fat and flourishing, though they are

4

forbidden by their rules to eat meat, eggs, or butter, and chiefly live on vegetables. "They eat," said John, "much onions!" Outside the building was a large square, with paved walks, for their hours of recreation, pleasantly shaded by rows of trees. Leaving the Monastery, we inspected Peter the Great's little cottage, in which he lived while superintending the laying out of the city. It contained much of his own handiwork, such as cabinets, etc., also his simple bed-room, and a big chapel, hung with Saints, where is the miraculous image of our Saviour, which accompanied him in all his battles. Outside the cottage was a boat made by his own hands. This was but the first instalment of the many relics of the great man which were shown us during our stay.

The sultriness of the day made so much walking and standing about very fatiguing, and I was thankful when John allowed us to leave, and gladly hailed his suggestion that we should take a quiet drive to one of the islands, though we little knew how lengthened the drive would prove. Leaning back in our carriage we obtained a little rest, and the air was now becoming cooler, while a refreshing breeze was wafted over from the Gulf of Finland.

After passing under leafy avenues of trees we came into a region of villa houses, ornamental wooden structures, with outside balconies, rustic verandas, flights of steps, and gardens, where were parties sitting peacefully at tea, enjoying the air in enviable laziness. Finally we reached an open space where many subsidiary streams seemed to unite in one channel, to be absorbed further away in the waters of the Baltic ; and then turning homewards once more, re-entered the streets, crossed the bridge of boats, and found ourselves again in the apparently interminable *Nevski Prospect*, and so home at 7.30 to our hotel, where we parted with our guide, with the customary fee of five roubles (about nine shillings), and with a friendly hope on his part of continuing our education with renewed vigour on the succeeding day.

I was however so thoroughly tired when I got up the next morning that I felt it necessary to keep my work within reasonable limits, and the great heat also prevented much rest at night. Our bed-room, though spacious, was very hot, and the incessant rumble of the streets seemed to go on all night, a noise greatly increased by the cobble-stone paving over which the droskies rattle. The streets are watered with hose from under-ground pipes, which partially keeps down the dust. We used to watch the rows of droskies, with horses and drivers standing patiently hour after hour in the hot sun, both man and beast often asleep, also the endless passing of little trains of drays containing earth, stones, &c., which is quite a feature in the Russian towns, the dray-horses,

like the carriage ones, having the wooden hoop
supporting the reins, like a sort of magnified
bearing-rein, and all without blinkers, their
great intelligent eyes peering from under shaggy
manes, and their luxuriant tails sweeping the
ground.   I would gladly have spent the day
quietly watching the street sights and writing
letters, which were always waiting to be written,
but L.'s sense of my unperformed duties outside
roused me at last, and we started again in a
carriage and pair, with John, to the School of
Mines, for all the distances in this immense city
seemed too long for a walk.   We were always
remarking on the vast proportions of everything,
the great squares, the huge, though not very
lofty houses, the parade grounds, big enough
for a small army, all which made the people
look dwarf-like indeed, while perambulating
the vast areas of the city precincts.

The Museum of the School of Mines contains
a wonderful mineralogical collection.   Among
the specimens we were shown a beryl worth
£5000, and a nugget of gold valued at £4000.
The gems were most beautiful, both cut and
uncut, also the pearls in their native oysters.
I think I never saw topazes of such beauty
and such exquisite shades of colour, some of
the palest etherial blue, others like smoky
crystals.

After this we were shown the Museum of
Imperial carriages.   Among the gorgeous equi-

pages was the sledge of Peter the Great, made by his own hands. It is like a small cab or sedan chair, the windows being composed of tiny panes of mica. At the back his travelling trunk is strapped, in which he carried his clothes and provisions. It was sent from Archangel by Alexander I, where Peter had left it when on a journey. We were allowed to get in, and sit on the actual seat once occupied by the old monarch, whereon he must have rested for many hundreds of weary miles.

We were told that the Royal carriages are all taken to Moscow, for the grand coronation processions.

We were also shown the wreck of the carriage in which the late Emperor was returning to his Palace when the fatal bomb-shell was hurled at him; a dreadful sight, with smashed panels and splashes of mud still on it.

After completing this survey we felt our dinner fairly earned, and on reaching our hotel we informed our guide that we should not require him again that day, an announcement he received sadly, but, as things turned out, he did eventually earn his five roubles, for L. in the afternoon finding her energies sufficiently restored, announced to me her intention of taking a walk by herself, and descending the stairs, she slipped softly out of the great front entrance. Alas! she had reckoned without her host. John, whom she had seen, as she thought, in a semi-comatose state in one corner, must have caught the echo of her step in his dreams, and before she could cross the street he was at her elbow. She had to confess on her return that she was really the gainer, for they had mounted the dome of St. Isaac's, and seen the whole panorama of the city, including the nine watch-towers, where are men stationed night and day, to give notice of fires, and to signal in which quarter of the city any may spring up; and as they returned they had actually seen one signalled, three balls notifying that it was near the third quarter. After this L. had been seized with a desire to pay a second visit to the Lady of Kazan, and had all but added, although in no spirit of devotion, her kiss of homage to the sacred picture, when deterred by the too numerous marks of lips devout. It was as she was recounting her adventures to me at supper, in one of the public rooms, that an exciting incident occurred. Two gentlemen were sitting at an adjoining table, who were evidently compatriots, but more than this, something in one of the voices was strangely familiar, and I also fancied I caught the sound of astronomical talk. We knew the tide of astronomers must be by this time setting in for the great Russian centres of observation, and one of these gentlemen proved to be an old friend, whom I had formerly met in America. His companion was a stranger to me, but mutual introductions took place.

My spirits were raised and cheered by the meeting, and knowing them to be *en route* for the same place as ourselves, I was able to send a message to my kind, but as yet unknown friend, Profsssor B., who had offered before I left England to find apartments for us at Kineshma. We did not meet them again during our short stay, but, as will be shown, we were destined to see more of each other before we left the country.

The next day was Sunday, our third since leaving home. It was intensely hot, and the usual thunder clouds were gathering, but no storm had as yet relieved the air, these clouds always passing away before evening.

We resisted John's persuasive endeavours to induce us to make a "Sunday excursion" to Peterhof, and walked alone to the Kazan, where we enjoyed the soft melodious singing. There were crowds of people, but even in their Sunday best, too close companionship was not desirable, but we sat for a few minutes on one of the side benches, watching the crowd of faces and the repeated genuflections with which the whole congregation seemed intensely occupied, some of them bowing till their foreheads touched the ground. The expression of their faces was serious and earnest, no idle glances strayed to the two foreign ladies, even from those close to us. The habits of the nation are devout. Even on the tram-cars we constantly saw business

men making the sign of the cross when passing
the shrine of any Saint.

The worshippers in the Cathedral were chiefly
of the lower class, but whether well dressed or
not, rich or poor, men or women, there was no
separation—all mingled together without distinction.   Now and again the Priests, in gorgeous vestments, would emerge from the Iconostas
to swing the censers of incense among the people,
the choir being always outside the screen.
After leaving the Cathedral we sat some time
in a garden not far from the hotel, where were
trees and evergreens.   A good many sallow-
faced children were playing there, in the charge
of nurses or mothers, and we noticed one old
woman sitting on the ground, busily reading
some religious book while she nursed a child.

We found the mid-day meal, whether as lunch
or early dinner, was generally ready about two.
It was served in this great hotel in a succession
of dining-halls, opening one out of the other,
which were pleasantly cool and quiet.   There
were many separate tables, and as soon as
the guests had made choice of one and were
seated, a waiter would bring the "carte," one
being provided in French for those ignorant
of Russian, but we often had to wait a long
time after making our selection.   We tried to
choose national dishes when it seemed possible,
but the ordinary fare was not unlike that provided in most foreign cities.   We frequently

had cold sturgeon, cut in thin slices, and looking
very much like veal ; it was deliciously tender,
and was served with vegetables and horse-
radish, and a slightly acid, creamy-looking sauce.
There were always melons or strawberries, with
cream and sugar, if desired. L.'s great difficulty was to get anything to drink, as she took
no stimulants, and all travellers are warned
against the Neva water, while perpetual "*tchai*"
was monotonous. She bethought herself of
lemons, and, after a battle with John's national
prejudices, had been permitted to buy some,
but "lemons are for tea," was his dictum.

After a quiet afternoon and a glass of tea in
our room, we went out for a stroll, with a pleasant sense of finding our way alone. The
evening effects were always most beautiful,
the light touching and bringing out the many
colours of red roofs and ochre-tinted houses.
The shops were open, the fruit-vendors about
in the streets, selling strawberries and pears,
or ripe gooseberries, while in places there
were stands, with tea in glasses. In the side
streets people were sitting on their door-steps,
the women wearing smarter handkerchiefs than
on week days, large silk squares, white, black,
or coloured ; the working men in picturesque
red blouses, sometimes with a black cloth waist-
coat over them. In spite of the heat we saw
winter jackets on some of the women. The
tram-cars and droskies were going about as usual.

Monday (August 1) was our last day at St. Petersburg; we were to start by the evening express for Moscow, and therefore wished to avoid a fatiguing day beforehand, also the intense heat continued, with cloudless skies. But we could not leave without seeing the Hermitage, the great Museum of the city. It is an immense building, with a fine portico, supported by ten colossal figures. Entering a large vestibule, where officers in red and gold livery were in waiting, one of whom took us in charge, we had to show our passport, and deposit dust cloaks, sun-shades, etc. The whole place seems associated with Catherine II, who was its founder, but it is also full of relics of Peter the Great.

Peter's gallery contains, beside an immense number of smaller objects, such as his watch, walking-stick, tools, telescope, etc., his own effigy, life-size, in wax, in the dress of the period, also the actual horse he rode, and his favourite dogs, stuffed, in a glass-case. We also saw curtains which were embroidered by his wife, for her own coronation day, cases of jewellery belonging to his wives, and a model of the little house he lived in at Rotterdam, where his landlady is honoured by appearing in the shape of a large doll.

One great feature of the Hermitage is the collection of pictures. There are galleries of all schools and all nations, England not excluded;

and really John won our admiration by the way
in which he had all information respecting the
pictures at his finger's ends.   He was also very
considerate in allowing us to linger by those
which interested us most.   As everywhere in
this mighty city, it was an *embarras de richesses*,
and though we passed through the whole suite
of rooms, it was impossible to do more than
inspect a very small part of their contents.
This was the last of our sight-seeing here, as
we found the crown jewels, including the Orloff
diamond, could not be seen without a special
order ; and so we drove back to the hotel for
the last time, through the great city, the vastness of which impressed us more, rather than
less, on each successive view, and never more
strongly than this morning.   Surely, we thought,
the *Nevski Prospect* is the ideal street.   As we
drove along, our eyes sought in vain for the end,
it still stretched on in hazy perspective before
us, on one side the colonnade of the Kazan, its
steps, as it was a fête day, crowded with people,
on the other the fine houses, the hotels, the
shops, while the street between was like nothing
I have ever seen before.   It was rather an
immense straight space than a street, the
breadth of which may be imagined from the fact
that five distinct passage-ways are included in
the ground between the buildings, three carriage
roads, two of them of wood, and one of cobble
stones, lying between the ample pavements for

foot-passengers. The blocks of the wooden
pavements have, we were told, to be renewed
in places every year, and the cobble-stoned
roads every two years, the terrible winter making
havoc of everything, though it was difficult to
believe any weather could affect such severe
and uncompromising things as those dreadful
round stones, which must certainly ensure a
comfortable and permanent income to the boot
makers of Russia.

In the afternoon we had our packing and
other matters to attend to, photographs were to
be bought, and L. was desirous of laying in a
stock of fruit for the journey. The assiduous
John, who accompanied her, imparted his decided opinions on this matter. " I should like
some pears," said L. " No," he replied, with a
look of disapproval, " No, Miss, no pears, they
are not right, cherries are for a journey." L.
felt so convicted of the betrayal of very plebeian
tastes that she humbly acquiesced at once, and
the basket of delicious fruit with which we
solaced ourselves in the train seemed to prove
the wisdom of her acquiescence.

We fortified ourselves for the journey by a
good dinner, good, even including some very
unappetising looking soup, of a greenish ditch-
water appearance, in which boiled eggs were
floating ; and then we started for the station,
accompanied by our guide, and most useful he
was in the hurry and bustle. Our tickets taken

at the hotel, for a sleeping-coupé, had to be produced. For these there is an extra charge of five roubles, the fare in an ordinary carriage from St. Petersburg to Moscow being fifty-five roubles. Unfortunately the carriage in which our lot was cast was at the very end of the train, which caused more motion, and the couches being back to the engine, added to the heat. To this we were fairly inured by this time, as our rooms at the hotel had been hot to an almost unbearable degree, a fact which, combined with other unwelcome inhabitants, had been one cause which had driven us away sooner than would have been really necessary. These Russian sleeping-carriages have a passage on one side, not, as in many European ones, in the middle, but other arrangements are similar. They are perhaps one degree better than a berth for rest. The train started at 8.30, and we rejoiced in the long daylight, as we now, for the first time, felt ourselves travelling in Russia, and before we turned in for the night we had tolerably possessed ourselves of the character of the country, which varied little hour after hour, *verst* after *verst*. The wooded parts might have been Sweden, but these were far less continuous than in that country, and a wide expanse of bare, desolate plains more usual. Sometimes we saw herds of cattle collected in groups, as though pasture was not too plentiful, with their attendant herdsmen,

man and beast becoming, as the darkness deepened, only visible as moving shadows, just
enough to keep up some faint semblance of
the life of an inhabited country. There were
indeed villages from time to time, which seemed
to rise suddenly out of the dim earth, themselves
hardly distinguishable from the dusky stretch of
the plains ; but I was agreeably surprised to
see that they were superior to the mere mud
huts I had been led to expect, resembling rather
the low rough chalêts one sees in groups on
the higher Swiss pastures. Over this monotonous country shone out presently the brilliant
colours of the sunset, a deep, lovely glow, with
not a cloud to catch its brightness, thus all the
more in harmony with the flat expanse below,
a glow which lingered long as a ruddy or golden
light on the far horizon. Then came a vision
of the silvery moon, now low in the heavens,
which glittered through and between the dark
stretches of the pine-woods, while in the west
Arcturus hung out his red lamp, just as I had
often seen him in my English home. Real darkness was of short duration, a little broken sleep,
and the dawn was upon us, almost a repetition of
the sunset, and revealing a landscape unaltered
in character by the miles we had traversed.
There had been several stoppages during the
night—there was one of ten minutes for supper,
and again in the morning at Klin, for breakfast,
where hot tea and coffee were never more

welcome. At the end of the Klin platform was
a really pretty little garden, laid out in neat
flower-beds, between paths gravelled with broken brick ; we noticed many of our old-fashioned
annuals, and, in a small, stone vase, filled with
water, birds were taking their morning bath,
such a refreshing sight that I had quite a difficulty in inducing L. to return to our carriage
in time.

CHAPTER V.

MOSCOW

The approach to Moscow is uninteresting. There is no view of the city, hardly any suburban district, no noticeable buildings, until the train is actually gliding into the station. All that met our eyes gave only a general impression of a dusty earth and a glaring heaven, without even a tree to add its grateful shade.

As we alighted from the carriage we were taken possession of by an elderly man, of melancholy aspect and subdued manners, who appeared to have singled us out by intuition as two beings still more unhappy, because unable to speak for themselves. His name we afterwards learned was Parlour, and he was one of the accredited *valêts de place* of the Slavianski Bazaar whither we were bound. "A truly appropriate name for a spokesman," said L. as we followed him. The omnibus which conveyed us to the hotel had four horses abreast, but neither from the windows of this vehicle did anything special meet our gaze, if I except a glimpse of the Chinese wall. We caught sight of English advertisements in some shop-windows certainly, as we drove along, but I regret to say without any patriotic rapture. Things of Russia were our heart's desire. Well, we had but to wait a little. The master of the hotel received us with every courteous attention, even assuring us that if Parlour should not prove what we wished, there were others who could be engaged with perfect ease. He conducted us himself to our bed-room, but on our entreating that it might, of all things, be cool and quiet, selected another with a sitting-room attached, and here, with the prospect of a week's tarriance before us, we gladly established ourselves.

Our three windows were very high, but had seats to which it was possible to climb. The outlook thus obtained was of a large courtyard through which vehicles occasionally passed, but otherwise quiet. Above its walls of pale red ochre, pierced with the hotel windows, sloped up a deep iron roof of a darker shade, the pink and red combining in great harmony. Beyond were visible other roofs, at various angles, of the same tone of colour, and peopling these ruddy eaves crowds of the soft grey pigeons of Moscow, a city of wings, as we soon found. It was by this time nearly one, and we decided to try and get some early dinner, but the *salle à manger* was far to seek, for after descending to the entrance hall we had to traverse several long corridors before reaching the great dining-room. It was an

5

immense place with ornamental alcoves or
recesses on the side-walls framing in the inevitable mirrors, but its speciality was a large stone-
basin in the centre containing living fresh-water
fish, chiefly the delicate sterlet of the Volga which
are highly esteemed. In this the fountain splashed
constantly, with refreshing sound, and here every-
one might actually choose his dinner. We saw
people doing so, and the fish they had selected
being lifted from the water. The sterlet are
ugly fish, like a finny eel. We tasted them
afterwards in soup, but their richness makes it
needful to indulge in them in moderation.

The universal "icon" or sacred picture, with
its lamp pendant before it, was suspended at
the end of the room, our own apartments being
similarly provided, though without the lighted
lamp. A number of little tables were arranged
about and down the sides of this great hall, and
the whole place had a cool and quiet effect, as
also had the brown holland suits which many of
the gentlemen were wearing.

Of the afternoon of this day I remember little,
except the great refreshment of some glasses of
tea in our own room, my memory-cells for the
rest of the day being retained and occupied for
ever by the evening that followed, to me the
cream and crown of all our evenings in Moscow.

To begin with its prosaic elements, it was the
occasion of our first drive in a drosky, which
was ordered for us by the *maitre d'hotel*, who

also gave the driver his directions. The space allotted for the seat, which faces the horse immediately behind the driver, is small for two persons. Our first impression when seated thereon, was of surprise, that in comparison with our expectations we were so little jolted, but I think that probably this sensation was only pushed aside by one more paramount, that of the necessity of holding on. The first impulse of a stranger is to clutch his companion and prepare for the worst, but this soon passes off with custom, and even on this our first experience, after having been conveyed in a rapid manner— whether a few yards or whether a street or two I would not venture to affirm—we became comfortable in our minds and quite able to look about us—and to look about us was a wonderful thing. The past slipped from us like a dream, but the sights to which it gave place were hardly less dreamlike. We were passing under great archways—we were in a great square— around us were domes and cupolas, lofty and solemn buildings, whether churches, palaces, or fortresses we hardly knew, and while our eyes were busy, suddenly our ears were caught and entranced by the breaking forth all around us of a chorus of bells. On the right and left, from one church spire after another, from belfrys near and others farther away, came this marvellous, musical, silvery chiming, such as we had never heard or imagined before, a peal soft and sadly sweet, yet with an indescribable joy in its melody. We were in fact in the Kremlin without knowing it, and the day was the eve of the Empress's birthday, for which the bells were ringing, and amid this sweet sound we were descending to another archway leading back again to the city, which from this elevation lay stretched out before us. At this hour everything lay bathed in the light of the setting sun, and as this light touched and was caught by one coloured roof or building after another, it produced an effect indescribably beautiful. All the roofs and houses of Moscow are painted in a great many different shades of soft green. There are also numerous other tints, the red tone of walls or tiles, blues, greys, besides the white and golden domes peculiar to Russia, and when all these were softened and blended in the mellow hazy light of approaching sunset, the effect must be seen to be fully realized. The sky was clear, and in the cool air hundreds of swifts were careering about, wheeling round the walls of the Kremlin as if they too revelled in the music. The bells continued at intervals, and when we reached the Boulevards a band was playing, and here our driver stopped of his own accord. Throughout our drive he had, silent and motionless, directed our flying course without demur or hesitation, doubtless following the orders he had received with the patient obedience of the born Moujik, his broad, padded back conveying to us a happy sense of ballast and security.

The next morning, (August 3rd) we engaged the melancholy Parlour and two droskies to take us to some of the churches, and thus made a further acquaintance with the Kremlin, and were able to realize its character more clearly. It is, as it were, a city within the city, containing besides the churches and other religious buildings of which I will speak later, palaces and arsenals, but no streets, no shops, no hotels. It may thus be said to be the high place of Moscow, literally from its elevation on a hill, figuratively because devotion, grandeur and autocratic power here find abodes prepared for them, but not the humbler needs or lower occupations of men. Outside, under the shadow of its mighty walls, cluster numerous places of merchandise, looking all the smaller in comparison with the lofty buildings which overlook them, many indeed being nothing but wooden sheds placed in rows round the open space immediately outside the fortress, the road which actually adjoins the walls being shaded by rows of trees. Each of the five entrance-gates is surmounted by a tower, through which you pass by a dark, roughly paved passage. Many of these gates are very picturesque, and not less so the bell-towers, for in Russia, as in Italy, it is most usual to have the bells separated from the church they belong to, an arrangement the more necessary

in Russia on account of the domed form of
roof prevalent in the sacred edifices.

Among these many turreted gateways is
one specially sacred, which is guarded by a
gendarme whose duty it is to see that all passing through it uncover and cross themselves.
We once kept a watch on Parlour, who seemed
inclined to shirk this ceremony, but he speedily
conformed at the approach of other passengers.

Passing in, we first came to the Church or
Cathedral of the Assumption, but to-day being
the Empress's birthday it was so crowded with
worshippers that we decided to leave our
inspection of it until another time, and passed
on to St. Michael the Archangel's, which contains the tombs of the Metropolitans, all railed
in and upholstered with red velvet, a singular
sight. The walls are covered with frescoes of
saints and saintly deeds from floor to roof, and
the preponderance of gold and of golden ban-
ners produces an effect of magnificence greater
even than in the churches of St. Petersburg.
The Cathedral of the Annunciation is the third
great church in the Kremlin, and as the Czars
are crowned in the Cathedral of the Assumption and buried in that of the Archangel Michael,
so in this they are baptized and married. Its
sacristy contains its own special relics, and it
has its own special history, and its stories of
John the Terrible, which meet one everywhere.
It was in this church the French stabled their

horses in 1812. While thus touching on the
churches of Moscow, for it can be but this, I
must not forget the Cathedral of St. Basil the
Beatified, which stands just outside the Holy
Gate of the Kremlin wall. It is a grotesquely
irregular building, having eleven domes, each
differing from the others in shape and colouring,
but all inclining to a bulbous, onion-like form, the
central one, which rises high above the others,
surmounting a tower, the architecture of which
has rather an Indian than a European character.
The interior of this church is equally strange.
L. remarked, that, if so profane a suggestion
were permissible, it was eminently adapted for
a game of hide-and-seek, for the whole space
inside is intersected with alcoves, with pillared
ways, and arched and winding aisles, where one
may seem to lose oneself, only to emerge in
the same starting place—the whole being also a
maze of brilliant, though rather coarse orna-
mentation, not in pictures, but in wall decoration.

The great heat continuing, we deferred our
sight-seeing the next day until the evening, thus
losing our chance of the best carriages, as
Parlour sadly informed us, but obtaining one
with two lovely black horses with wild manes
and tails, driven, as is always the case, without
touch of whip, the long, looped up reins being
used instead. Our guide had evidently, as L. remarked, spent the intervening time since the
previous morning in mournfully considering our

ignorance, and had arrived at the conclusion
that it would be needful to begin our education
from quite the lower strata of life in Moscow.
He therefore proposed to take us a "simple
drive" into the suburbs "to see the nationality"
and "how they enjoy themselves in Russia."

The festal place to which, with this intent, we
were conducted through clouds of dust, was a
wood of silver birches, growing out of the arid
dusty soil of this flat, treeless district, and
providing shelter for a multitude of little tables,
each with its white cloth, *samovar* and glasses,
the ingredients of the "cup that cheers" being
brought by the visitors. The people were
slowly arriving in a long straggling stream of
pedestrians, looking dusty and tired enough,
but eminently contented, and in tune for the
national enjoyment when seated at their respective tables. The women were mostly
handkerchief-capped, though some had modern
hats or bonnets. Our driver, who had conveyed
us here at a rapid rate, now slackened his pace,
and having slowly made the tour of the wood
turned his horses home. Coloured glasses for
lamps, chiefly red, were being put up everywhere
on tall posts, in readiness for the evening
illuminations, and tri-coloured flags, the com-
mercial colours of Russia, were flying. At every
cross road were stationed gendarmes, mounted
or holding their horses. We met some equipages of the better class, on our way home, with

fine horses, these grander coachmen also without whips, but this evening there were no bells,
and, a singular coincidence, no circling swifts.
As we reached our hotel, thunder clouds were
gathering, according to their nightly custom.

Thursday (August 4th). The heat was very
distressing. We now knew our way unattended
to the Kremlin, and as L. and I started thither
alone that morning, the scorching heat of the
pavement seemed to penetrate our shoes. We
felt it a relief to be alone, to look hither and
thither as we felt inclined, and during our stay
of ten days in Moscow, these little rambling,
impromptu walks were among our pleasantest
experiences. We would sometimes take a
drosky, as to walk even a short distance over
the cobble-stones in the heat was exhausting,
and we had discovered that there were two
classes of these vehicles, the better ones, for
which a double charge was made, being cleaner
and altogether more desirable. Sometimes, in
the evening, after strolling round the Kremlin,
which always acted as a magnet to our wandering
feet, we would return by the river, the Moskva,
which partly encircles the city, and where the
evening reflections were often marvellously
beautiful. The sullen, copper-coloured clouds,
which often gathered at sunset, would give a
ruddy tinge to the water in places, or a red-tiled
warehouse would repeat itself, or the keel of a
boat, or the long white walls of the Foundling

Hospital, and we often remarked that these water-pictures, which are common to all countries, had their own special character here, partaking of the indescribable and fascinating charm of Moscow. Then, as we wound our way upwards again, there would be the little ruby stars of the "riza" lamps at some street corner, or under some dark archway, often surrounded by a group of worshippers, and in the half-light the persons we met seemed to us to gain something more than his or her everyday look—the eyes that mildly encountered the gaze of the two foreign ladies, seemed to hold something special and distinctive, but to me mournful, like eyes which hardly knew the light of a free country.

Of these street sights and experiences L. carried away perhaps a larger store than myself, as her walks were more frequent, by which I was also the gainer through her lively and graphic descriptions.

One afternoon she went to the Iberian Chapel. This is a little place hardly larger than an ordinary English china-pantry, containing a copy of the picture of the "Iberian Mother of God," which is more venerated than any other, the Emperor always dismounting when he enters Moscow, and praying before it. Of this sacred picture a second painted copy, in a loose frame, is hung in the chapel, to act as a substitute for the other, when this, as often happens, is carried to the houses of the votaries who from sickness,

or any other cause, are unable to visit the
chapel, the demands for this privilege being
sometimes so numerous that they can hardly be
complied with. L. noticed that the feet and
hands of the figure were black, and the paint quite
worn away by the kisses of the devout, and as she
stood in the chapel, which she did for some time,
she saw for herself how incessant these worshippers were. Ladies who had been shopping
would come in and kneel down, and touch the
ground with their foreheads by the side of some
poor woman or shock-headed youth in soiled
working garb, all, before leaving, kissing the feet
and hands of the picture ; and she was greatly
interested in a fine looking soldier in his grey
military cloak, who managed to secure a front
place, and who gazed at the painting with his
beautiful soft brown eyes full of adoration, his
lips moving all the time, and his hands constantly
making the sacred sign. L. could not get over
her surprise at the reality of this evident
worship of sacred pictures by intelligent men,
and we were told that in the army it is very
conspicuous, a priest with a movable chapel
accompanying every detachment.

On another occasion L. departing myster-
iously by herself, announced on her return that
she had relieved her sense of duty by doing the
old clothes market, feeling sure *my* sense of
duty was not sufficiently strong to support my
other senses under the ordeal. There, as she

passed rapidly along, any lingering among these commodities and their purchasers not being possible, she had seen a number of moujiks gathered round a little booth or shop, where a man was opening and holding out for inspection a number of second-hand sheepskin coats. Lads perambulated the place with other old garments hanging over their shoulders, while the old-stocking department seemed handed over to the women. She saw a group of three women and four men all assisting at the purchase of one second-hand suit, while mixed up with these old-clothes shops, were stalls laden with bread, with small cucumbers, with sausages, cheese, fruit, etc.—everything you would or would not expect to find in such a market. As she went by, the sellers would shout out, " Buy, Damski, buy," regardless of the inappropriate nature of their wares. There were some better shops near this market which I *did* visit, but of the shops of Moscow as a whole our impression was that they contained few articles specially national, and for this reason we did not find it easy to select many souvenirs for presents or for ourselves. The confectioners shops had one peculiarity common to other countries, the absence of that useful, common-place thing, a bun. Biscuits were also unknown, but there were sugary compounds, chocolate bon-bons, and such-like things offered in abundance.

About this time we had a little spell of wet weather, and thus learned something of the mud

and holes of Moscow streets. On these wet
days we used to watch the pigeons from our
windows, sheltering themselves under the eaves,
or sitting in rows on the cornices which often run
round the houses. We counted fifty in one row
on the opposite house, sitting closely packed together, other rows of possible fifties being
apparent further off.

We still resigned ourselves at times to
Parlour's services, and it was under his guidance
that we, one day, visited the Palace in the
Kremlin, a palace not much used as a dwelling
by the Royal family, except at the time of a
coronation. A great part of it is comparatively
modern, fire after fire having consumed the
earlier buildings, but we were shown one interesting old part where were the rooms used by
Alexis, Peter the Great's father, and the old
throne, like a well-worn arm-chair, with steps to
mount by.

Time had toned down the original brilliant
colouring of rooms and furniture, till the
fading hues blended together in a sort of
subdued richness, which we thought pleasing
and appropriate. In this we differed from
Parlour, who kept repeating at every turn.
"Very gaudy, very gaudy, is'nt it, miss?" The
carpet in the banqueting room was, we were
told, entirely the work of the nuns of one
of the convents—cleaner nuns, we hoped, than
those we often saw standing at the church doors

with their books held out for *kopecks* by hands
which looked quite unfit to embroider anything
for royal feet to tread upon.

After seeing the council-chamber of the
Boyars and many apartments magnificent in
size, one requiring more than 4,000 candles to
light it, we walked round an outside gallery
which gave us a beautiful view of the city below,
with the river winding in and out, and, better
still, a breath of fresh air after the close stifling
rooms.   It was a scene we never tired of at
any hour of the day, though in the evening
the light touching the golden crosses which
surmount every church made them a more conspicuous, and very striking feature.   In the
court yard of the palace is a very little, old,
wooden church, originally placed in a small
wood, which once covered the hill on which the
Kremlin stands.   It is called the Church
of the Saviour in the Wood, and contains many
relics of St. Stephen of Perm, the first Christian
missionary martyr in Russia, in 1396.   We also
saw a place of terror in some stone steps outside
the palace where Ivan the Terrible did, or is
reported to have done, many horrible and
bloody deeds, the very Ivan who had the Acts
and Epistles translated into the mother tongue
and distributed among the people, so strangely
are opposites combined in character.

I have but touched on this morning's work,
and must, in like manner, only touch on another

of our expeditions, a visit to the Treasury, also
under Parlour's guidance, where we inspected
endless stands of fire-arms, military accoutre-
ments marvellously be-jewelled, chain-armour,
state carriages—not to speak of coronation
robes and crowns—and were once more sur-
feited with precious stones, the crown of the
Empress Anne alone containing, we were told,
2,536 diamonds. Among the thrones the
most interesting was one used by both Ivan
and Peter the Great, in which, when the
drapery was pushed aside, you could see the
aperture through which his sister Sophia used
to prompt the duller brain of Ivan. We could
but skim this immense collection of splendours
and curiosities, amongst which we again came
across the abounding relics of Peter the Great
which had so amused us at St. Petersburg,
including his enormous boots, which certainly
proved him to be great in more senses than one.
There was also, strange to say, a relic of the
hated conqueror Napoleon, his simple camp
bedstead having a place in the collection.
Whether it was the sight of this relic or not
which recalled that phantom of conquest to our
minds, I do not know, but it was in the afternoon after our visit to the Treasury that we
suddenly bethought us of the Sparrow Hills, the
place where Napoleon is said to have caught his
first sight of the Holy City, after his long forced
marches with his army, and where, turning

pale at the sight, he hissed out through his clenched teeth, "*Moscow, Moscow !*" We determined to go there, and to go without Parlour, so, chartering a carriage and pair, we boldly started by ourselves at 5 p.m., when the heat was a little decreasing.

The late rains seemed to have made but little impression on the streets of Moscow, which were once more becoming dusty, and the suburbs through which our road first lay had rather a dreary look. Yet there were pretty bits of life and bright bits of colour—among the former a house surrounded by a wooden fence with fresh, green grass inside, evidently a girls' school, where we saw the girls moving about in white tippets and aprons. There were little houses here and there, painted bright red and green, and larger houses with gardens, also outlying churches and monasteries, until leaving the city confines, and with them the cobble-stoned road, we rejoiced to find ourselves on soft ground. This rejoicing was of short duration, for the softness became mire, the mire mud, the mud a sea of brown, watery slime, like a ploughed field in time of floods. Ruts and holes were the only road, into which, however, our horses boldly plunged as a matter of course, our carriage swaying beautifully from side to side, working out its own track, ourselves momently expecting to end our trip in a mud bath. This state of things was in part accounted

for by a rail-road which appeared to be in process of construction, as we saw wooden sleepers
being laid down, and strings of small carts
passing backwards and forwards. Emerging
from this slough of despond, we presently
perceived houses appearing in the distance, and
in due time our coachman drew up before a
wide, open veranda, where tables were arranged
and the *samovars* in full swing. Of course we
alighted and ordered our "*tchai*" like the rest,
and then were indeed rewarded for our jolting
by a splendid view of Moscow on the horizon,
but of hills we saw no trace above or around us.
The slight ascent we had made in driving here
from the city, had, we supposed, in this land of
flatness gained the place its name.

Moscow *is* beautiful seen from a distance.
You can descry the river in its devious course,
embracing in its curves and twists the vast city,
with little rowing-boats moving lazily along,
and though it was so remote, with the shadows
on the banks clearly defined.

In the foreground were gardens and clumps
or copses of trees. As the sun sank lower, the
churches lighted up their spires and crosses as
usual, and the golden dome of the new cathedral
was the last to be illumined, and to hang out its
tiny ball of light. I feel I may be alluding too
often to these lovely evening effects, but they
are one of the greatest charms of Moscow. We
looked forward to them day after day, and felt

6

that even those prosaic souls at home and abroad, who never lift their eyes to a glorious sunset when its very glow is on their faces, must have awakened here to some sense of the exquisite beauty of the scene, and yet to me the sun never set quite without sadness, from a dim prophetic dread that one day now drawing so near might prove a day of shadow.

As to that day our prospect was in one sense most happily assured, for before this a very cordial letter had reached us from Professor B., offering us his and his wife's hospitality in their country-house, a kindness which would also insure to me the additional pleasure of the society of our two English astronomers, who were to be our fellow guests.

The day of our departure was already fixed, but there was one more excursion we felt we must not omit before leaving — that to the Troitsa Monastery.

To reach the Troitsa Monastery, you first go by rail to Sergiefoskaza, and this was a rather interesting part of the day's work. We accomplished our ticket-taking without help, and settled ourselves comfortably in the carriage. After leaving the city the line of rails passes through a wood of fine old fir-trees, something like Scotch firs, except that the trunks were a palish yellow rather than red. Some of these ancient trees were quite bowed down, and the little paths which wound in and out beneath

them, made us long to wander there   Further
on we came to villa houses, and to the old type
of pine woods with which we had become
familiar.   We saw English wild-flowers as we
passed along, hawk-weeds, knap-weeds, and
rag-wort ; there was one bank with a coating of
lichens, and tall grasses were abundant ; we also
noticed some leaves like a lily-of-the-valley.
Then came a long waiting in the refreshment-
room, which was crowded with pilgrims, as
indeed the train had been, though, happily, not
our carriage.   Pilgrims flock daily to the Holy
Place, and many of those we saw had travelled
a great distance on foot, some of them probably
hundreds of miles, and veritable pilgrims they
looked, covered with dust and mud, their clothes
ragged, their legs swathed in bandages, and
their queer, loose shoes having the appearance
of old felt-baskets.   With wallet on back and
staff in hand, they might have stepped out of
some mediæval picture.

The better class, who had travelled by train,
returned with their baskets and handkerchiefs
full of little loaves, which had been blessed at
the shrine.   Having fortified ourselves with
some rolls, we took a carriage to the Monastery,
which we reached in about a quarter of an hour.

It is a curious place like a walled town or
village.   Within the enclosure are no less than
ten churches an ecclesiastical university, *ateliers*,
where sacred pictures are painted by the monks,

and workshops for the metal framing of "rizas" and "ikons." The encircling walls are white, and pierced with small windows. Outside the walls are crowded numberless little shops, booths and wooden sheds, and the whole place swarms with beggars and pilgrims.

The Holy Well of St. Sergius, the patron saint—in appearance a plain basin of water beneath a gaudily painted canopy—was in the middle of the square, and dates from 1342. There are many legends connected with the place of miraculous appearances of the Blessed Virgin to the Saint, besides histories of his successful intervention in military affairs, and of the rapid growth of the fraternity, which at one time boasted of thirty monasteries, and numbered 160,000 serfs as its vassals. It has been besieged many times, and an altar is shown where Peter the Great was once successfully hidden with his mother.

We went into one church where service was going on, and saw much glitter of gold, and of red lamps burning in the distant sanctuary, but the frescoes on the walls struck us as of coarse, inferior character. We should have liked to go the whole round, and to have visited the workshops, but we had no guide, and I confess to having shrunk from much close contact with the crowd of inodorous pilgrims, but when I had retreated and was struggling to make a few outline sketches outside, L. ran back to try and get a sight of the shrine. She first returned to the refectory where the monks were actually at dinner, and into which we had furtively peeped, and had seen them seated at three different tables, provided with three different kinds of bread, white, brown, and black, with wooden spoons for the soup, while the reader stood in the middle of the room, apparently reading from a Bible, during the meal. This room L. boldly entered, and stood waiting her opportunity. Presently a good-natured looking man passed her and she endeavoured, but vainly, to make known her wishes. He listened politely, and then spoke to a priest, who, bowing, motioned to her to stand a little aside, and then the good natured man hurried away, and in about five or ten minutes, during which L. did feel rather uncomfortable, returned with a superior-looking monk whose "*Parlez vous Francais, Mademoiselle?*" was a joyful sound to her.

Under his guidance she was shown the shrine and other sacred things. There was an agate with the appearance of a crucifix and a kneeling monk in its cloudy crystal, which in tones of the deepest veneration, her guide informed her was "natural, not made by human hands."

Returning to the station, we had our dinner. There was cold sturgeon, cheese, and some queer-looking patties, somewhat resembling dough-nuts, which contained a small quantity of hot-spiced meat and were very savoury. We

sat down outside to wait for our train.   People
were coming and going, chiefly pilgrims we
supposed, some carried boxes, and the number
of pillows tucked under arms was amusing. One
girl had a white cat wrapped up in a shawl, and
there were a good many children sitting down
having tea in the waiting room, all of the lower
class. Arriving at Moscow, we found the station
yard crowded with carriages, those with pairs in
front, the droskies behind. We bargained to be
taken to the Slavianski Bazaar for a rouble, and
felt very proud at having accomplished our day
so well. Happily it had been fine; but rain
came on again after our return, the sound of
which was not enlivening.

The dropping of the rain reminds me of
another sound which travellers in Russia are
likely to remember, and which arises from the
peculiar nature of the washstands. In these the
basin is filled from a reservoir fitted above at
the back part of the stand, and worked by a
pedal close to the ground. If this is pressed too
violently a rush of water is liable to deluge the
clothes as well as hands, which is a misfortune,
as the supply, which seems from the force of its
exit as if it were inexhaustible, is soon spent,
but whether used lavishly or not, the result is a
continual dropping, which, especially in the
"dead waste and middle of the night," is a
sound which refuses to be ignored, and which
L. declared was a perpetual commentary on the

contentious woman of Solomon. I must add
that these Russian basins are often made with
an aperture at the bottom covered with wire, so
that on touching the pedal the water from above
only passes over your hands into a can beneath--
an arrangement which certainly requires a re-
arrangement of one's ideas.

Moscow sounds and sights were soon to end
for us now. On Friday (August 12th,) we woke
to the consciousness of our last day, but the only
through train to Kineshma being a night one,
we could not leave till evening. I had managed
to break the spring of my watch, and had rather
an amusing experience this morning in a visit to
an optician. Our dumb-show not proving
equal to the occasion, the master of the shop
asked me in Russian whether I could speak
French, which did not much mend matters, and
I should have had to retreat ignominiously had
not another gentleman come to the rescue,
through whose help I explained what I wanted,
and obtained the loan of a big, silver watch,
without any glass over its face, to use until our
return journey.

There is one peculiarity in the streets both
of St. Petersburg and Moscow which I have not
yet mentioned. The sign-boards over each
shop, with name of owner and other information, are hung very high, at about an equal level,
and being profusely gilded they give the effect
of waving lines of gold along the streets, at

sunset, or at any other time when the light is in a direction to touch them, which is really very striking. We noticed it this evening as we drove to the station, an occasion on which we narrowly missed our train altogether, for the people at the hotel were much in the dark as to the hour of starting, and it was only by risking the earliest supposition (7 p.m.) that we got off that night. This was the only occasion on which our invaluable Bradshaw failed us. As a rule, we found its information correct, including the fares, which greatly lessened the difficulty of taking tickets unassisted. The station must be more than two miles from the Slavianski, but, thanks to the four strong horses of the hotel omnibus, we dashed up just in time, the "Bazaar slave" helping us with our tickets, and with the extra charge of two roubles for overweight of luggage. Fifty-five lbs. is allowed for first-class passengers. Hurrying on to the platform L. cried out, " Damski" (Ladies) to the porter, but apparently there was no ladies'-carriage, for he conducted us to a large saloon where were fourteen easy-chairs arranged to let down into small couches for the night. We congratulated ourselves on being alone, but on re-entering the train after alighting for supper, we found a burly Russian making himself comfortable for the night. In an astonishingly short time he was snoring on his great square pillow, indeed so rapidly did he enter the land of nod, that we really thought he was feigning sleep, and L. had vague terrors lest he should be on the watch for our small properties, but when one disturbance after another produced not the faintest impression on his sonorous slumbers, we decided it was true that a Russian can sleep any and everywhere, and devoutly wished that we could be Russian too, or that we had also provided square pillows, for the night seemed endlessly long, and the jarring of the train always gave me a headache. We reached Vladimir about the middle of the night, and L. crept out and had a turn on the platform, but saw nothing but moon and stars, and two old women sitting by their little stall in the light of a lamp, eating gooseberries to keep themselves awake, and again at Nowski, in the early dawn, when she saw a dismal waiting-room, full of people looking as if they had been there all night, all with pillows, or bundles to serve as such.

Between six and seven we got some breakfast; and by this time the snoring gentleman had been forcibly roused up by the conductor, who had paid us several visits to snuff the candles—large ones in little boxes placed high up in the carriages. We had kept one window open all night, and had not been troubled with dust, owing to the recent rains, and, hour after hour, we had watched the train of sparks rushing by, like fireworks, from the burning wood with which the engines are fed.

The country on which we opened our eyes in the morning was pretty; picturesque villages of small wooden cabins were scattered about; we saw some cows, but rarely sheep, and there was corn stacked in places in little square sheaves, and hay still in cocks : and now my big watch pointed to 10.15, and the train slackening speed, we found ourselves with beating hearts gliding into the station of Kineshma.

CHAPTER VI.

POGOST

To find oneself more than 2000 miles from home, on the platform of a little station in the heart of Russia, tired and confused with a night journey, with no knowledge of the language, and in the presence of a perfect stranger, is an experience that would, I think, make even a self-confident person feel somewhat shy and uncomfortable, and to pass from this desolate and troubled state of mind in a moment to one of grateful ease and comfort, could only be the result of exceptional courtesy and kindness of reception ; and it was just this that happened to my companion and myself as we alighted from the train at the little station of Kineshma. A tall gentleman of stately presence, and wearing a long cloak, was evidently looking out for us, and when my hand met his friendly grasp, and my eyes his kind, encouraging look, I felt all my fears melt away, and knew in a moment

that of all our pleasant experiences the pleasantest was yet to come.   And indeed my fears
had been many, and had been by no means
confined to the fact of adventuring myself
amongst strangers.   This gentleman, my host,
was one of the leading astronomers of Russia,
holding a high post in Moscow, and, in addition
to this, his two English guests, whom we were
now to meet, and whom he had invited for the
occasion of the eclipse, as representatives of the
Royal Astronomical Society of England, were
also so infinitely above me in scientific status
that I could but feel the amount of my own
self-acquired knowledge small indeed in comparison, and quite inadequate to place me as a
co-worker on their higher level.   It was in this
case also that my diffidence was again met with
such kind and courteous encouragement that I
was immediately set at my ease.

Our first introduction over and our small
parcels collected, we found Professor B.'s
caléche, with its four horses harnessed abreast,
waiting outside the station.   Into this we were
quickly handed, and were soon plunging through
the rough, sandy roads of curious, straggling,
untidy-looking Kineshma.   The carriage lurched
from side to side with jerks and bounds, sur-
passing even our experience of the Sparrow
Hills.   It was impossible to help laughing ;
we all laughed, the Professor laughed delight-
fully as we were thus shaken out of all stiffness

and propriety. "*Horrible !*" he cried, again
and again, consoling us in French after every
successive shaking, and then we reached the
Volga, a mighty, rushing stream, on whose banks
the town nestles, and where warehouses and
factories edge the shore.

Here we were transferred bodily to a raft, the
dear horses trotting on to it as though it were
an every-day event. It was a busy scene ;
people were waiting about, and the raft was
already encumbered with country people, their
bags and baggage ; and then a small rough-
looking steam-tug puffed away with us, winding
in and out of the huge craft which were floating
on all hands. On the other side was a second
landing-stage of rough planks, and having surmounted this, the carriage struggled up a steep
road, through a fir-wood, a road if anything worse
than the last, and a heavy pull for the horses,
and then, through a gap in the trees, we caught
sight of a large, rambling building. This
was Pogost, our host's summer residence, a
delightful, romantic house, like no other we had
ever seen, full of wide, wooden staircases and
roomy landings, and unexpected corners ; with
bare, polished floors, and with a long balcony,
overlooking a cool garden, where grass and
flowers were mingled together, and which contained a small piece of water, with a bathing-
house attached. Beyond the garden was a little
belt of wood of birch and willow.

Our kind host took us upstairs himself, and
introduced us to his family, which consisted of
four ladies. It was some time before we could
quite understand which was our hostess, but we
soon found Madame B. was the lady who could
speak a little English, the other ladies only
speaking French. They all received us most
kindly, and it was impossible to feel we were
not really welcome. We had already shaken
hands with our two English friends, who were
standing outside when we arrived, and had
caught a passing glimpse of telescopes and
instruments in a grassy court-yard, but it was
some time before we could arrange our ideas, or
fully realize where we were, or the geography
of the house. Our cheerful bed-room had three
large windows, with heavy curtains which could
be drawn up like blinds, the two larger facing
the road, and a smaller one giving a glimpse of
the great plains beyond. Just below, within
reach of our hands, acacia bushes hung out
their ripe pods. But the room which we fell
in love with at once, and which was our
favourite to the end, was one known as the
" balcony room," which was reached from the
upper landing. It was a sitting-room, carpetless of course, and provided with a number of
comfortable arm-chairs and a round centre-table.
The French windows were always open, so that
the balcony seemed quite a part of the room,
and some of our party were always passing in

and out. It was a place where one could watch the sky or speculate on the weather, or stand to breathe the sweet, cool air.

Dejeuner was served soon after our arrival, in the large dining-room with its long table, its straight wall-lines broken by evergreens, in wooden boxes—among them a very tall India-rubber plant—and the windows of which were festooned with the green leaves and tendrils of a climber, new to us, and which I think they called the Chinese vine. It had small pointed leaves, something like beech leaves, and reached to the ceiling.

We were not able to change our dusty travelling dresses, as the boxes had not arrived, but neither now nor during our stay was the matter of dress any anxiety to us. Ladies of the upper class in Russia always dress most simply in the country, and we envied our lady hostess her cool washing prints, and admired the perfect ease enjoyed by all the family in these troublesome matters. Even in the evening no elaborate toilet is expected, the ladies sometimes only adding a white shawl to their morning costume, so that any trepidation we had felt as to the limitations of our travelling wardrobe, in the prospect of this unexpected visit, proved needless. We found ourselves more inclined to feel self-satisfied than otherwise, from the evident interest evinced by our new friends in our dress and appearance, and in all our English belongings, even to

the smallest item. On the occasion of this
our first meal in the house Madame's sister
presided, and my place was by her side. She
talked French fluently, but it was very difficult
to me to understand her, and we were afterwards
told that a pronunciation often obtains in Russia
among cultivated people which would puzzle
a stranger. The Professor's seat was not at
the foot but in the middle of the table, but he
really seemed to be at every one's side, his
genial merriment and kind attentions touching
every one in turn; but L., whose seat was nearer
to him, naturally came in for more personal talk
than I did, and had enough to do at times to
hold her own against his witty sallies, a struggle
she thoroughly enjoyed.

Our English friends, who were now quite
habitués of the house, helped us greatly in
our conversational straits, both of them being
good linguists, and Dr C. having, during his
short stay in the country, picked up a good deal
of Russian. The talk during lunch, which was
a fair specimen of all our pleasant meals, was
a mixture of Russian, French, German and
English, Madame bravely addressing us in our
own tongue, in which her English guests informed her she was "much improved already."

Two men-servants, Eugene and Ivan, waited.
The former we learned, to our surprise, was
nearly 70, while Ivan was quite a boy. "Too old
and too young!" Madame laughingly complained.

The four dogs of the house, Hercules, Jupiter,
Arabe, and Zirchock, also occasionally put in
an appearance, stalking familiarly in when so
inclined, big, noble-looking creatures, the first
of whom still bore the scars of his last winter's
skirmishes with wolves. Immediately after
dejeuner the samovar is brought in, preserved
strawberries, with little glass dishes to eat out of,
being handed round with the tea. We passed
the afternoon quietly in our room, dinner being
from 6.30 to 7. Unhappily for my comfort a
headache, caused by the jarring of the train,
had by the evening arrived at a point when it
could not be concealed, so I had to retire for
the night before dinner was over.

The next day (August 14th) was Sunday,
a fact rather difficult to realize in this strange
land and place. Presenting ourselves in the
dining-room about ten, we found the gentlemen,
but no ladies, present. It was not their habit
to join us then, a servant pouring out the coffee
into large cups for the gentlemen, and smaller
ones for us, adding sugar as a matter of course.
The table is not formally arranged for this early
meal, as with us; the milk, bread, eggs, &c.,
being placed just anyhow. It was generally, to
my delight, a time when a little astronomical
talk went on, our astronomers being too polite
to "talk shop" before the whole party, but
of course at all times hopes and fears for the
weather on the 19th could not be kept out of

7

the conversation.   It was not a very encouraging
outlook even at this time, the settled hot weather
having quite broken up.   This morning, as I
stood at the window, soft fleecy clouds were
constantly passing over a sky of rainy blue,
and the little lake showed grey through the
birch trees.

The two English gentlemen had been at
work here for more than a week, and had
had a hard time of it in the great heat, Father
P. narrowly escaping a sun-stroke at one time.
They told us they had often begun work at six,
and worked ten or twelve hours a day, with
men to help, the Professor giving instructions
in Russian.   Before this their labours had been
great, in having to convey their instruments
for such an immense distance, instruments
weighing collectively many tons, and very valu-
able ones, lent for the occasion by different
Observatories, and some of which had made
long journeys before.

After coffee my telescope was brought out and
examined.   It had made its long journey safely,
and was fixed for me on the balcony, and
quickly adjusted, and then we went to see the
" camp," now in order and ready for rehearsals.
Professor B. had three splendid telescopes there,
the largest with spectroscopes attached, and all
were protected, at least partially, by a wooden
shed.   The great refractors and the other
smaller instruments—cameras, driving clocks

&c., made quite an imposing array, the large
instruments being quite imbedded in the earth.
The grassy court known as "the camp" was
within view of the road, and we often saw peasants who were passing stop and gaze curiously
at it, with somewhat awe-stricken faces. They
were aware something was about to happen in
the over-world, and although the eclipse had
really on this occasion been prepared for by the
Government, by the circulation of hundreds of
little explanatory notices, to allay terror and
insure assistance, it was inevitable that a dread
of something unknown and supernatural should
still be felt among people so imperfectly educated.

In the afternoon Professor B. took us all a
drive. To our eyes there seemed to be no road
at all, but to the horses, of which there were
three to day, this seemed no hindrance. They
danced along, little encumbered by harness,
through the corn-fields, we six in the *char-à-
banc* keeping our equilibrium as best we could.
At one time a horse would be quite above the
others on a bank, or perhaps quite below them
in a ditch, but the equine minds preserved their
balance, and really the horses do seem to have
a happy time of it, which must be strengthening
to the nerves of those behind them, as no acci-
dent ever appears to happen.

The corn-fields were delicious, with their
waving grain, and wide expanses, populous with
birds rising out of the ridges, and edged with

wild-flowers, including some pretty, delicate pinks and blue larkspurs, and the air of the country was most reviving after our city life, the one thing to mar our pleasure being the threatening sky.

Monday was again cloudy, and L. returning home from an early dip in the bathing-house attached to the lake, pronounced the day "chilly," and the water "tadpoly," but we were too bright and happy in our surroundings to take this much to heart.   Letters from home had also arrived, although the receipt and sending off of such seemed to us a very casual and uncertain sort of business, but our friends did their best for us, and before our letters were posted the Professor always added some mysterious hieroglyphics, to ensure their right despatch.   Every day we learned to know these kind friends better.   Madame B.'s English was a great help, and the extent of her attainments surprised us.   One evening she translated a short Russian telegram into English at sight, and a longer article on the coming eclipse into French.   She knew most of our best authors through translations, and was eager to take advantage of this opportunity for improving herself, so that we had our talks in English every day.

Our little travelling manual was a constant source of amusement to the others, and our futile attempts to master the pronunciation of

Russian words. When I heard L. unusually merry I always knew the manual was to the fore.

One of the ladies, Mademoiselle Lisoch, was a kind of lay-sister in good works. We found she was the helper and adviser of the country round in matters of health, and that she passed hours every day in making up simple medicines, etc., for the peasants. After this whenever we saw any of these poor people passing near the house we always set them down as Mademoisell Lisoch's patients. When the rest of the family return to Moscow one of Madame B.'s sisters still resides in this lonely house during the cold, dreary winter, when the roads are impassable with snow, and the whining howl of the wolves echoes nightly in the adjacent forests, thus keeping alive a little burning fire of help and comfort for the scattered dwellers on her brother's property.

This afternoon, which, for a wonder, was fine, we had a walk in these same woods, treading the mossy or pine-needled paths, where the sun-beams slant so brightly through the red stems, or on the banks where the pinks and yellow clover grow, with a sense of perfect safety ; and yet, hidden somewhere in these recesses, were these savage creatures, wolves and bears, who had left their mark on poor Hercules' wolf-like coat. How softly the wind soughed through the pine branches this

afternoon as we strolled on, passing great ants'
nests, which reminded us of some in our woods
at home, but the ground was wretchedly sloppy
with the late rains.

There was a stranger at Pogost to-day, a
little wiry, keen-eyed man, with long, heavy
black hair, who we thought seemed to watch us
with much interest.   It came out presently that
he was a press-correspondent from Kazan, who
had been inspecting the instruments and other
preparations, with a view to eclipse-articles.
He was present at *déjeuner*, and I noticed a
running fire of jokes going on among L.'s
near neighbours at her expense, while the
grave stranger continued to level keen, furtive
glances at us.   L. told me afterwards that our
mischievous host had amused himself by mis-
leading the poor man as to our respective per-
sonalities.   " He told him," said L., who was
convulsed with laughter, " that you were the
wife of the English Astronomer Royal, and
that I was an American young lady, assistant
at the Cincinnati Observatory, and constantly
employed in calculating lunar inequalities !"
The Professor had also tormented her all lunch
time by declaring that her every act was being
noted down, such as what she eat, what she
refused, &c.   Our Astronomers were also made
to pose as celebrities other than themselves ;
but whether we all figured in this guise in the
Russian press we never knew.   Besides these

occasional jokes L. had much good-natured raillery to endure on the score of her water drinking—which water, by the way, was, at Dr C.'s request, always boiled—and especially when she would persist in touching glasses, with water only, for a toast.

The correspondent had disappeared by dinner-time, but there was a lady-visitor, and visitors often dropped in in this easy fashion, partly, we surmised, for a glimpse of the strangers. After dinner—at the end of which finger-glasses were always brought in, with a little glass inside them, containing warm water for rinsing the mouth—coffee would be served in the drawing-room, a room more bare-looking and not so snug, to our thinking, as the balcony-room, for it does not seem the habit to put out ornaments or books in such rooms. We brought out our small store of sketches and photographs this evening for the amusement of the party, also some of my sister's needlework, and quite regretted we had not more English things to exhibit.

The total absence of books or papers on the tables surprised us. It was probably one sign of the police supervision, which makes a visit of search possible and permitted in private houses at any hour of the day or night.

The weather all this week was really distressing. An exasperating fine drizzle would come down for hours, which would clear off for awhile only to come back again.   The so-called roads grew worse daily, and little wonder, for the district produces no stone, the only road-mending consisting in throwing down a cart-load of bricks or of logs when the ruts have become quite past endurance.

This morning (Tuesday) we watched a cart go by, drawn by a horse by nature white, but the colour of whose coat was literally concealed by sandy mud splashes.   The road just outside the entrance court was a constant source of amusement to us from our bed-room window, the queerest vehicles, generally hooded, passing not infrequently, containing passengers to our eyes very queer also, besides the daily diligence or *char-a-banc* from Kineshma.

But for the rapidly approaching 19th, the weather would not have troubled us much.   As it was, L. often declared she was thankful not to be an astronomer, and indeed we were all getting restless and anxious, but to-day my spirits were greatly raised through the kind offer made me by my host, through Father P., of the use of one of his beautiful instruments for the eclipse.   While rejoicing over this L. came running in, in excited trepidation, to tell me that she had been retained by Father P. as head-assistant and slide-shifter at his camera during the eclipse, " so I am to have my share of anxiety with a vengeance," she said, " and a harder two minutes' work I shall never have had in my life!"

She had her first rehearsal in the afternoon, when it cleared for a time. " What had you to do ?" I enquired. " I have to get three slides in and out," she said, " which have to be wrapped in black bags. I have to open shutters, to run round to another camera at the other side. One shade has to be exposed a minute, another half-a-minute ; and this leaves me just half-a-minute to ·do all the manipulations. Then I have to shout the time for some man to put down, and he has in turn to shout back in French. All which, if you please, will have to be done by the light of a tallow-candle in the dark. ' *Don't get flustered !*' said Father P. to me, and I should like to know how I am to help it. To crown all, I have to stand and jump on and off a pile of ricketty packing-cases, and, *whatever* I do, I am not to give the telescopes the *slightest* jar! I told Father P. I was sure to come down and bring the whole apparatus down with me ! Well," concluded L., " I hope the six saints in our bed-room will give me their blessing, for I shall never need it more !"

The next day, Wednesday, (Aug 17) was a lovely morning, after the rain, at last. The other inmates of the house being all occupied— the astronomers, as usual, taking photographs or developing plates in the dark-house, L. helping, and the ladies invisible—I took the opportunity of slipping away for a solitary ramble.

Providing myself with a camp-stool, I ran down
the wide, wooden staircase leading from the
upper story, and passing through the porch and
the grassy court-yard before mentioned as "the
camp," which is the only carriage-approach to
the house, and where the precious instruments
were all standing, I turned to the left, through
the opening in the trees and shrubs which
served for a gate, and found myself on the high
road to Kineshma.   The ruts were full of water,
and I had to pick my way as best I could, con-
gratulating myself that a pair of goloshes had
not been left out of my box.   Leaving the
house and its surroundings behind, I emerged
into the open plain, covered with stunted crops
of wheat and other cereals, and stretching as
far as the eye could reach, although the more
distant views were intercepted by birch-woods,
which, to my thinking, gave rather a dreary
appearance to the landscape.   I followed one
of the many narrow foot-tracks which led
through the golden corn, now dead-ripe, and
waving in the wind.   Truly, as they say, the
labourers must be few, to leave it so long un-
garnered.   Many peasants, men and women,
were coming and going, who never failed to
salute me with the usual respectful obeisance.
Their tall figures stood out in bold relief against
the horizon, their red blouses being conspicuous
afar off.   I looked in vain for any subject for
my pencil, anything but a coloured sketch, and

that very much idealized, being out of the question, so I passed on, and at last beginning to grow weary, I turned aside from the main road and made for a group of young saplings, under whose scanty shade I planted my camp stool, and, not to return quite empty-handed, I made a rough sketch of the Professor's house, which was visible in the distance, with its row of Corinthian pillars on one side, showing white among the surrounding shrubs. I was completely alone, the corn having been cut and gathered up into sheaves. There was no sound of bird or beast, only an occasional cry from the distance broke the perfect stillness. Presently, on looking up, I saw thunder clouds gathering, so putting my things quickly together, I hastened for shelter to some larger trees not far off. They proved, when I reached them, to be part of a double avenue of birch-trees, which skirted the high-road all the way to Kostroma, planted, we were told, by Alexander I. Here I waited until the shower was over. The domes of a church were visible at some distance ; it was the one nearest to Pogost, from which the parish, or its Russian equivalent, derives its name, but it was too far off for me to reach on foot. I was so well pleased with my morning that in the afternoon, which was clear and still, I explored another corn-field in the opposite direction. Here I saw a huge barn, with open doors, half filled with sheaves, which some men inside were beating on a wooden shelf. The structure was entirely of wood, with openings at the top for the escape of the smoke, when fires are lighted to dry the grain. Several pigeons were feeding contentedly and unharmed on the floor. In the evening we noticed smoke ascending from the barn.

Thursday morning (Aug. 18) L. was at early rehearsal again, but Father P. kindly spared time to draw out some circles for me, that I might make rapid sketches of the corona, with a view to estimating its extent, as compared with the sun's disc, an important work, as the fainter parts of the corona cannot be photographed. We all felt terribly anxious. Dr C. appeared with his throat muffled in a white silk handkerchief, but we all decided his sore-throat was nothing but nerves, though he stoutly denied this, or that he possessed any, and accepted my proffered camphor, as for an orthodox cold. Father P. alone retained his philosophic calm, but then failure would not mean to so veteran an observer what it meant to us. Our spirits rose as the day advanced; the sun shone out, and a drive was arranged for the afternoon. We had the *char-à-banc*, with two horses, and the Professor and Mrs B. and we four visitors were packed in, back to back. This time one of the horses not only got into the ditch but fell on his flank, but

almost before we knew of it had picked himself up again, and was going on as if nothing
had happened, but we thought we could hear
the coachman remonstrating under his breath
with his "little father" for his carelessness!
The village was a romantic one, especially for
this flat country. It lay along a sort of gully,
or small ravine, overlooking the Volga. The
pine-trees, scattered about among the wooden
châlets, gave it quite a Swiss appearance, and
some of the little wooden houses were very
pretty. It was the Festival of the Transfigura-
tion, and the people were all in holyday garb.
They stood round me as I was sketching, but
gently and respectfully, not pressing or pushing
rudely, as one often experiences. Almost all
the women and children wore red cotton print-
dresses, some with loose jackets over them.
Even the tiny children had a handkerchief tied
over their little heads. I longed for a rapid
artistic touch to put in some of these groups
in colour, but had no time to do it justice even
in pencil.

We noticed tall poles standing close to some
of the châlets and rising above the roof, with
wooden boxes fixed to the top, and in some
cases with a bunch of dead birch-leaves festooned over them, probably, as in Sweden, for
the sake of ornament. These were houses for
the larks on their first arrival in the spring, when
the snow is yet on the ground, and when they
are welcomed as the harbingers of summer.

Madame B. took us into one of the cottages, explaining to the people that we were English, and would like to see a Russian peasant's home. The two little rooms were clean and neatly arranged, though very bare of furniture, but of course the sacred picture was there.

On our way back we met two English gentlemen, who had stopped at Kineshma in the course of their travels, for the eclipse, and who remained to dinner. Two assistants for the next morning also arrived, Russian students, of rather military appearance, whom our astronomers evidently thought had come too late to be of much use, not having gone through L.'s severe training. By this time she was able to get through all her work, to Father P.'s satisfaction, a few seconds under the two minutes, these seconds being reserved for "shouting" the time. How well I remember the evening before the great day, in the balcony-room ; the gentlemen stepping in and out to get the right time by the stars ; the lamp burning over our heads ; the attempts at general topics always coming back to the one—the strong feeling of a strong united interest. They told us a gendarme was to be stationed outside to keep off curious intruders, and the dogs were to be shut up. Father P. hoped the bells would not be ringing, the sweet Kineshma bells, almost if not quite equal to those of Moscow—bells whose like I shall never hear again—and so we said good night once more in hope.

It does not need the unfolding of these pages to tell any who have kindly followed our track thus far with interest what the history of the 19th really was. All the world knows that astronomers, with few and rare exceptions, were to meet with a grievous disappointment, how grievous few outside the astronomical world can fully realize ; but I must tell my story to the end, the more so that there are few grievous things without compensations, and we had many.

L. and I were stirring soon after 4.30 to inspect the weather. It was unpromising enough, the whole sky above the horizon being cloudy. By 5.30 we were all up and dressed. As I went downstairs I met Father P., who seemed hopeful, as there was a patch of blue sky and a gleam of sunshine. The ladies were hurrying on the early coffee. We were all at our posts in good time, matting laid down, pencils sharpened, candles in readiness, L. mounted on her boxes ; the new assistants were hastily drilled, the clock set going, with its double row of figures, to mark what would remain of totality.

Still the light drift-clouds kept passing over constantly, with only one short break ; and it was through this drift that we saw the moon surely, steadily, beginning to pass over, to eat away, as it were, the sun's disc. Standing motionless by my telescope I kept glancing at the horizon, which continued clear, but the clearness never reached the upper heaven, and

as totality approached it became worse rather
than better.   Darkness was now upon us, a
peculiar gloom, which gave the grass a ghastly
hue, and I found I could no longer direct the
telescope to the sun.   I had an opera-glass too,
but nothing seemed of any use.   Suddenly,
when the two minutes and a half had nearly
elapsed, there was a general, involuntary excla-
mation.   For a second or so we had a view of
the coronal light, making it look almost like an
annular eclipse, and a glimpse of the rose-
coloured prominences, as they are called, on
about one-fifth of the circumference, but it was
over almost before we had realized it, and I
never knew *how* I saw it, whether through a
glass or with my naked eye.   Father P. exposed his plate instantly, and indeed some of
them had been going on photographing in the
darkness, as a kind of forlorn hope.   Another
second and a voice said—"The eclipse is over!"

As the light returned it showed a very dis-
consolate group.   We could hardly look in
each other's faces, and I am sure no one just
then could find voice to speak.   Almost the
first words we heard came from Dr C.'s Scotch
assistant, who muttered just audibly " *We'll get
home,*" a remark which seemed to chime in fitly
with our feelings.   L. said afterwards it was
like a death-bed, the more so as Dr C. and his
helper began at once to encase some of the
instruments in the black, coffin-like boxes.   We

felt so sorry for each other, it almost stifled
personal feeling, and so dreadfully sorry for our
kind host and hostess, for they had so lived in
our hopes, and indeed Madame B. was quite ill
with disappointment, and could not appear at
dejêuner. By that time we had pulled ourselves
together a little, and made shift to be cheerful,
but though we tried to talk on other subjects
we could get up no interest in anything else at
all, so Father P. occupied our minds by recounting other failures, equally impossible to be
escaped by human effort, but I think just then
that we all echoed our host's remark that "an
astronomer's life is hard;" and yet—so quickly
does hope and resolution rebound in the heart—
one of my first thoughts and questions was,
"When and where will the next total eclipse
be?" but this no one could answer in a very
encouraging manner. And then letters had to
be written, not a cheerful task. One of mine
was to the Secretary of the Liverpool Astronomical Society, with which I have the honour of
being connected, and to whom, in the first place,
a full account was due. This too, with flagging
pen, had to be accomplished.

Before dinner our friends arranged a drive,
and the diversion of thought did us all good.
This time it was to pay a call at a neighbouring
house, a pretty villa, with wide veranda, overlooking the Volga, and with steep paths running
down to the sandy shore. L. went down to the

8

river, but I sat listening to, rather than joining in, the lively conversation which went on for nearly an hour. I felt too tired out, mentally and bodily, to take my part in it at all creditably, and then we drove back to dear Pogost again, for our last evening. And here, before good-byes are finally said, let me, in case these pages should ever meet the eyes of our kind entertainers, assure them here, that, keen as was our disappointment, we carried away with us a feeling keener still, the grateful sense of warm hospitality received in a strange land, the happy memory of a pleasant week in their pleasant home.

CHAPTER VII.

NIJNI NOVGOROD, AND SMOLENSK

During our stay at Pogost we had mentioned our idea of seeing the great fair at Nijni Novgorod before leaving Russia, but it was thought hardly suitable for ladies, and we were told it had lost much of its distinctive character, however, when our plans were finally discussed we were advised to make the Volga part of our return journey, and in this way, to get a glimpse of Nijni, where we could take another line of rail back to Moscow. We therefore left Pogost on the morning of the 20th, in time to catch an American steamer which would pass Kineshma about 2.30 in the afternoon. Madame B. kindly came with us in the caléche to the landing-stage, and there, with mutual expressions of regret, we parted, after she had seen us safely on board. The river looked lovely, its wide surface glittered in the sun ; on a little promontory not far off a troop of horses were standing, as if watching the bustle. We found the steamer a good one, and our cabin was clean and roomy, with an attentive stewardess, who provided us with sheets on our paying for them, these articles never being brought as a matter of course. These Volga steamers are not places for rest, if only from the constant use of the discordant steam-whistles, either when stopping at any place, or when passing another boat ; but we should have been sorry to leave the country without seeing the great river. We mounted to the upper-deck in the evening—a sort of roof to the lower ones—and there sat enjoying the air and watching the sunset light on the thunder-clouds, the reflections on the shore, the stars above, and those below, of the red and yellow lamps which are placed to indicate the channel. It is only three Russian feet, about three English yards, deep, and boats frequently stick in the mud on either side. Wooden poles are provided for such emergencies, a sort of wooden leg, to help the ship to walk along. Before the daylight went, L. had been an excursion round the ship under the guidance of a gentleman whom we had met at Pogost, and who introduced her to the captain, but as neither of them could speak more than a word or two of German they finally handed her over to the first-mate, who knew a little English. He showed her the machinery, telling her the quantity of wood consumed in so many hours, and other items of information, some of which were not very reassuring, for the danger of fire in these boats seems very great, from the quantity of wood carried, and it is possible for a ship to be entirely destroyed by fire when not far from the banks, for which reason, candles are always used on board instead of lamps, as being safer.

Women are largely employed in loading the boats with this fuel. We saw them hard at work by lamp-light, bare-footed, and carrying heavy burdens of it on their backs which they threw into the hold in the most light-hearted manner possible. One effect of this burning wood is very pretty, the train of sparks passing over our heads like showers of rockets.

Another gentleman befriended us by helping us to order dinner, beginning with sterlet soup. He told us, to our aggravation, that he had seen the eclipse quite well, only half-an-hour's distance from Kineshma. He was standing near a wood, and the great Russian hares were so terrified by the darkness, that they came running out, and crouched close to him, as if for protection.

It was very windy as we neared Nijni the next morning, so much so, that we could hardly get through our tea on deck without the cups—they were very small ones—being blown away.

We reached the town between seven and eight. Its appearance from the river is very picturesque, part being on high ground, and the river itself being a perfect forest of masts, boats, and craft of every description, which had brought people and merchandise to the fair.

We waited for the crowd, of what I may call steerage-passengers, to go on shore first, all bearing great bundles, and all showing that *absence* of cleanliness which seems the sign of godliness here. We had watched them on board and seen them eating bread taken from disgustingly dirty bundles, crossing themselves devoutly before eating. The fourth-class passengers only pay the value of about two shillings for a voyage of 400 miles, providing their own food.

When our turn came to disembark, one of our kind helpers saw us into some droskies, directing the driver to the railway station, where we were to wait for the night train to Moscow, himself going with us part of the way. He seemed hardly to like to leave us unprotected, and mentioned his address at Nijni, that we might come to him if in any need of help. This was another of the many instances in which, while in Russia, we met with every possible help given in the kindest way. The station was large and dreary, and crowded with people from the fair going off by some train. We waited for the confusion to subside, and then had a second breakfast, but when it came to requiring a drosky to see the place, we felt that our difficulties were beginning, until, remembering our late friend had mentioned the telegraph office as a place where some language other than Russ was possible, we betook ourselves there,

and securing the good offices of a kindly clerk,
were safely started on our sight-seeing. We
first drove to a height overlooking the Volga.
There we saw the two mighty rivers, the Volga
and the Oka, pursuing their course through
an immeasurable expanse of open, alluvial plain
which seemed to have no limit or end. We
alighted from our vehicle and sat under
some trees in a kind of dusty park, the shade
being refreshing after the still dustier streets;
but the "*tchai*" which we ordered and expected to find good, was the worst we had
tasted in Russia. All the same it refreshed
us a little; and I took a hasty sketch, and
then we climbed into our drosky again, which
had been waiting for us, and drove back through
the fair, and down a hill so steep and fearfully
rough that even the horse objected to it, and we
ourselves flatly refused to be driven. The glory
of the fair must indeed be departed, if it ever had
any. To us it seemed nothing but a combination
of crowds, dust, and evil odours.

The dirty streets were in some places encumbered by men lying asleep on the ground,
whom we had to avoid treading on; and to
drive through narrow lanes of shops whose chief
commodity is skins, dressed or undressed, is a
trial to more senses than one.

The chief novelty to us was the number of
Tartars in the fair, some in squalid garments,
others gaily dressed in green silk robes. These
form a large proportion of the merchants, and

were to be seen walking about with their goods hanging on their arms.

As we were returning we saw a train of poor prisoners on their way to Siberia. They looked miserable in mind and body, and were guarded by soldiers with drawn swords. Though it passed quickly before our eyes it was a sight difficult to forget.

I was thankful to get back to the station, dreary as it was, and told L. nothing would induce me to go out again; but she was brave enough to make a second excursion by herself later on in the day, returning with a pair of Russian boots of thick white felt peeping out from under her cloak, to the evident amusement of the station porters. How she managed her bargaining I cannot imagine. She told me she was at one time quite surrounded by Tartars offering her skins, she having, in an evil moment, attempted to ask some question relative to these undesirable articles. However, she returned alive and rather excited by her adventures.

Our whole experience that day was as opposed as it could possibly be to any previous experience of Sunday, the only token of the day being that the woman in the waiting-room, taking advantage of a period of comparative quiet, was reading a little Gospel of St. Mark. The hours dragged heavily, our dinner was a risky business, as we could only guess what the eatables were. I espied some mushrooms, but found they were pickled, and very sour, but there was cheese, like Gruyére, to be had, and I managed to buy a bottle of Médoc.

Our train was due to start for Moscow at nine, and we were thoroughly glad, after thirteen hours here, to settle ourselves into a carriage, but it did not add to our comfort for this night journey, that instead of one snoring gentleman we had the company of twelve, and were the only ladies; but we did manage to get some sleep all the same, although our travelling companions *would* smoke a little, and also positively objected to an open window.

When the dawn was creeping on, and our journey in view of its end, we stopped at a station where a welcome and unwonted refreshment was offered. On the platform were benches with rows of basins, soap, and towels, for washing, with women standing behind. We lost no time in availing ourselves of this unexpected benefit. The soap was many-coloured, and some of the towels prettily embroidered. This extempore lavatory was provided at several stations, and our gentlemen had all rushed out at a previous one, before we were really aware of the opportunity.

We reached Moscow at about 9.30 on the 22nd, and established ourselves at the Slavianski. It seemed quite home-like and natural, and for the day we fell again into our old habits, L. going

out in the afternoon to reclaim my watch, and
both of us turning, almost without thought, in
the evening, in the direction of the Kremlin, for
one more view of the Holy City, one more walk
round the walls, one more listening to the
melodious bells.   Only a single small adventure
varied the routine.   We had become so bold
that we had taken our sketch-books, and I had
just finished an outline of an archway and had
put up my pencils, and L. was still at work,
when a gendarme came solemnly up and made
signs to her, with great gravity, that it was not
permitted. Of course she immediately acquiesced
and so we were not locked up or otherwise
reported to head quarters, neither were our
books confiscated, a possibility which had
caused us a momentary tremor of dread.

The next morning we started for Smolensk.
This was a little bit of obstinacy and adventure
on our part, for we had been distinctly advised
at Pogost *not* to go there, but we wished to
break the long journey to Warsaw, and had an
idea it would give us a glimpse of one of the
more distinctly national towns, as indeed it
did.   We left Moscow at 11.30 for a long
day in the train, but this time were fortunate
in having a nice ladies'-carriage to ourselves.

It was fine all day, and we often longed to be
out of the hurrying train, and wandering at our
leisure along the soft green turf under the
shade of the trees, which were almost all silver-

birch.   When cut and stacked the lovely
glistening bark of these trees gives the effect of
a slight covering of snow.

During this day's journey we saw more arable
land than we had before.   Barefooted women
were engaged in harrowing, two harrows being
drawn by a single horse which is placed
between them. We also noticed a great number
of black sheep. We did not reach Smolensk
until after 10 p.m., and it required some courage
to leave the safe shelter of our carriage and go
out into the darkness in a strange place, not
even knowing the name of a single hotel, or
able to direct the driver if we did.   All the
disparaging opinions which had been expressed
to us about the place came back very vividly
then, and when we heard a party of Americans,
who left the train to get some supper, remark with
somewhat scornful emphasis "Those people are
getting out *here*," misgivings did certainly arise in
our minds.   It was too late then to give way to
them, and once more befriended by a kindly ticket
clerk, who explained our enquiries to a porter,
we were despatched by the latter in two droskies,
each with a pair of small horses.   The droskies
were dirty, and one at least of the drivers shady-
looking, and we were of necessity parted from
each other just when we would most gladly
have been together.   Off we rattled into the
gloom, over two railway-bridges, bright with
red and green lamps, and up a sandy hill.
Lights were twinkling here and there, but we

hardly seemed for some time to be approaching
any town, and when we reached some flickering
gas-lamps which might indicate streets, the place
seemed asleep or deserted.   Shortly after our
drivers turned into a street with no lamps at all,
the only light visible being in upper windows,
where the inhabitants were apparently going to
bed, and here L. who led the way, confessed to
having felt just a little creepy, a horrible story of
a lady who was murdered by a drosky-driver,
coming into her mind, so that she kept anxiously
looking back to be sure that I was following.
We could say nothing and do nothing, however,
but hold on and keep a tight grasp on the brass
labels, which the friendly porter had given us, as
our only safeguard.   The Great Bear was
blazing overhead with other familiar stars, and
the night air was reviving after our journey.

Presently we crossed another bridge, a wooden
one, over the Dneiper, which rolled dark and
gloomy under the night, and then mounted
another hill with a sharp descent on one side of
it—but no hotel appeared, indeed it seemed the
last place to look for one.   At length the domes
of some church became faintly perceptible
before us, and then more gas-lamps came into
view, and lights under trees; we heard the music
of a band, and then suddenly, to our immense
relief, our drivers drew up at a door.   It proved
to be the Hotel de l'Europe, but it spoke no
European language but Russian, and we had to

fall back on signs and our manual for bed and supper, making signs for a second bed in the large stuffy room allotted us, and for coffee, as we were too tired for any regular meal.

In the morning light the place looked more ordinary, and less like a foreground for dark adventures, and we were glad we had broken our journey at a place so essentially Russian. After breakfast—when we vainly tried to get an omelette, receiving poached eggs instead; and after yielding up our passports rather unwillingly to a man of doubtful appearance, we set out on a voyage of discovery. The town is hilly, which adds to its picturesqueness, the Dneiper running right through it and doubling under several bridges. It has a battlemented wall of brick of great antiquity, with castellated towers. This has been broken down for some distance and the spaces filled in with newer buildings, and with an archway. We wandered on, over bridges, and through curious streets where the houses in some places looked black with age, and whose battered stones seemed to bear witness to the scenes of tumult and carnage which the old place has witnessed, and then all at once we came upon a market filled with country men and women. It was an opportunity for seeing the real Russian costumes, such as we had not had before, the men in their sheepskins, the women with thick, whitish coats of some material between felt and flannel, or in long jackets of deep blue, over short pink cotton skirts with handkerchiefs on their heads ; some of them with bandaged legs and feet encased in flat, basket-like shoes, costumes which looked more suited for winter than summer, and which had the appearance of needing a good scrubbing in the river. I wished it had been possible to draw a few figures in colour, and I did make the attempt from the door of a harness-maker, whither we went to buy a few of the little bells used for horses.

The horses here seemed even less troubled with harness than in other places, many of those in the country-carts having no bits at all, only a rope knotted round their noses.

Rain came on after a time, and we had to take a drosky to go back to the hotel. Early dinner was a curious meal there, as it was laid under a sort of low, open balcony connected with a garden, where was a theatre with coloured lamps hung round it. It was rather a dreary dining-room in the drizzling rain, and the soup looked very much as if it had been ladled up from a stagnant pond, and the floating, whole eggs and pieces of meat added afterwards.

We spent the afternoon quietly, our train not leaving Smolensk till 10.15, when we chartered two droskies to take us to the station, and again made the transit in the eerie dark, but this time with firmer nerves.

Settling in with the prospect of a day and
night in the train, we were glad once more
to have a carriage to ourselves. Sleep came to us
more kindly to-night, and we did not rouse ourselves until about 5.30, when a porter announced
a stoppage, and we got some coffee, but found no
washing apparatus provided. The country
looked much the same. There were the same
birch and pine-forests, with occasional clearings
and stacks of wood, but few signs of human life
or dwellings. The day was cloudy till afternoon,
with a little rain, and quite cool. We stopped
at many places, and again found the dough-nut
like patties, with a little minced meat inside
them, served with the meat at the "buffets."
They are quite a peculiarity of these countries,
and are very good, though rather too rich for
a journey. Before reaching Brest Litovski,
where we entered Poland and changed carriages,
we crossed a river and passed through much
flooded country. There we heard the cry of
water-birds, of which we saw large flocks. They
somewhat resembled herons, but had great
white wings tipped with black, and orange legs,
and they seemed to be quite without fear of man,
standing close to people who were at work.
There were also flocks of geese and of the
hooded-crows.

We reached Warsaw about 8.30 (August
25th). It had been a trying day and night for
poor L. who was out of sorts and greatly in

need of the quiet it was impossible to get in a train, so that we were both thankful when this, our longest spell of railway travelling, was accomplished, though to leave our carriage was to plunge into a turmoil of noise and confusion — touters for the hotels shouting, a crowd of carriages, many with coloured lamps, moving about, and all the din of arrival in a large city. Our omnibus turned out to be only a close-carriage, which we shared with two not very polite gentlemen, and, packed closely into this, we drove to our hotel. The city looked vast and mysterious in the light of the gas-lamps, and full of stir and movement.

As we crossed the broad Vistula a troop of cavalry were passing over the bridge in their white tunics and scarlet trousers; fine men and powerful horses, all fully equipped as if for a march. Our hotel, another Hotel de l'Europe, was a large and noisy one. English was spoken here, and we were at once pounced upon by a man, who proved to be a "commissionaire," who demanded our passport, and also the names of our parents, a piece of information which had not been required before. This man showed us our room, and seemed inclined to take possess-ion of us at once, but L. was too unwell the next morning for any sight-seeing. I had a turn by myself, down the street and across the iron bridge, and did a little shopping, and in the evening we managed a drive, not very successfully, as the coachman took us through the suburbs.

The Vistula looked pretty in the hazy evening light, with distant clumps of trees standing up against the sky, and we afterwards went to some gardens near the hotel, where we saw our first chestnut-trees since leaving England, and also ribbon flower-beds.

Throughout the city, all names, advertisements, etc., are both in Russian and Polish, and our sympathies were strongly enlisted for the poor conquered nation. It was a feeling that seemed in the very air. We looked with sadness at the Poniatowski Palace, now confiscated to Russia. The very language is confiscated. All official correspondence must be in Russian, and it is imperative in all the public offices and schools.

The next morning (Saturday, August 27th) had to us the feeling of a last day, for we were to leave Warsaw in the afternoon for Berlin, and it seemed like a return to more beaten tracks. We were obliged to accept the help of the "commissionaire" for once, but his loud, strident voice and over-bearing manners were very disagreeable, also the amount of praise which he demanded as a right for every object of interest which he showed us. Our train left at 3.40, and it quickly bore us into country more civilized and cultivated than we had seen of late. The stations had their acacia-trees instead of the untidy shrubs of Russia, and the great open plains showed carefully-tended crops, the corn being mostly carried, while large patches of yellow lupins were still in full blossom.

Our passport was demanded at Alexandrovna, the frontier.   It was a large station with good waiting-rooms, where for the first time during our journey mashed potatoes were served with the cutlets.

The luggage was examined at a station further on, but our boxes were not even opened, and then another night came down upon us with the imperfect rest of a railway carriage, by this time a familiar experience, but in the present case, of a carriage so shaky that our efforts to sleep were almost unavailing.

Our short acquaintance with Berlin, which we reached at 6 a.m. I do not profess to recount, for though a glimpse of it was very interesting to us, and though we made the best of our limited time, we were not able to gain more than a passing impression.   We were on the home track now, and a feeling of longing for return to old England grew with the passing hour, and gave us no wish to linger.   The sights and sounds of continental life were to be before our eyes, and in our ears for yet a few more days, but they ceased to hold us with the charm of our first starting, or with the novelty of the stranger and further-off lands which we had visited.

Cologne and Brussels belonged to old experiences.   They were little more than places

of rest to us now on our journey, just stages of
the way, whose last scene was the steamer
from Calais to Dover.   A stormy scene, for we
crossed on a day so rough that even some
experienced sailors hesitated to start.

Wrapt in every available cloak, shawl, or
tarpaulin, we kept on deck, though the waves
came right over the vessel's side, making the
planks one wash of sea-water.   It was in the
midst of this strife of the elements, of the
whistling winds and tossing sea, and through
the mental confusion and misery of sea-sickness,
that the white cliffs of our own country shone
out on us as a welcome beacon-star at last.

We had laid up a store of memories for all
our lives—there was time enough to recall and
revive them.   We had followed in the track of
a shadow, and had proved there was abundant
light behind—the gleam of pleasurable excite-
ments—the vision of things new and beautiful—
the glow of accomplished resolution, and of
surmounted difficulties—the warm shining of
new friendships—yes, and the dawn of hope that
would not be wholly crushed, of better success
yet to come!

Printed in the United States
By Bookmasters